Internet of Things and Big Data Analysis: Recent Trends and Challenges

Ali Al-Sabbagh and Ruaa Alsabah

Florida Institute of Technology, Melbourne, FL, USA

United Scholars
Publications

United Scholars Publications, USA
Our distinguished editorial team thrives to produce and publish
superior quality books and journals.
Visit our website for the current list of publications
www.unitedscholars.net

Editors: **Ali Al-Sabbagh and Ruaa Alsabah**

Florida Institute of Technology, Melbourne, FL, USA

Published by United Scholars Publications, USA
Copyright © 2016 United Scholars Publications
www.unitedscholars.net
info@unitedscholars.net

ISBN-13: 978-0692809921

ISBN-10: 0692809929

Disclaimer
The Publisher and the Editor\Author hold no liability for
incidental or consequential injuries or damages caused by the
information contained in this publication.

UNITED
SCHOLARS
PUBLICATIONS

Preface

In recent years, there has been a wide interest in smart objects that can be connected to the Internet for the sake of sharing and hence creating enormous and innovative services to the world. Such materials are utilized in smart homes and mobile health. A very similar development is experienced in industrial applications, such as in smart grids, efficient transportation and logistics. The Internet of Things (IoT) consists of a large number of objects with intensive connections, which allow these smart objects to be sensed and controlled remotely across an efficient network. Such an innovation is the next best thing and will be utilized in different aspects of life. This book presents the fundamentals of IoT including Big data analysis and the 5G trend with their most important challenges, features, and explanation of how they will drive IoT to build smart cities and smart objects. The objective of this book is to analyze the data in order to make better decisions and create a clustering of complex knowledge in an IoT big data situation. However, there are still several challenging issues that demand to be addressed before the present IoT ideas becomes broadly accepted. Big Data is related to growing data sets that are of large and complex volume with multiple autonomous sources. Data comes from many different sources, i.e. web apps, mobile apps, sensor networks and many others. These are row data which need to be collected, categorized, stored, and analyzed in order to get an insight on their content so that they are presented in an efficient way to the end user. The current rapid growth of the Internet of Things (IoT) in various commercial and non-commercial sectors has led to the deposition of large-scale IoT data. Similar to other technological applications handling sensitive information, it is necessary for IoT applications to be developed and deployed with highly secure mechanisms to protect the data and services provided by these technologies. Existing security approaches need to be improved to allow adaptation for requirements of the IoT platforms. These are platforms that lack resources and computational power. On the other hand, IoT devices can generate an enormous amount of data which becomes increasingly difficult to process using the limited computational and storage capabilities of these devices. To tackle this problem, a notion of a novel technology called Cloud of Things is emerging. However, to harness the full potential of this new paradigm, different security and privacy issues need to be properly analysed. Furthermore, Information and Communications Technologies are transforming at a rapid rate driven by emerging economies, smart cities, and certain key technologies including big data and a broadband of wireless technologies. To meet these requirements, a number of new network architecture designs and concepts are greatly in need. In particular, 5G will drive massive innovations due to the fact that it is currently underdeveloped. Thus, it is an opportunity to enable high capacity fast rate, reliability, low latency and connectivity through extending the heterogeneous networks and exploiting a new frequency land. The key developments will involve M2M, V2V, D2D and IoT infrastructure that facilitate the advancement of future smart objects, buildings, and cities. It is worth noting that many of these applications will open up partially through the use of the 5G trend. This book is organized into nine chapters. Its target is highlighting the trend and the future vision of the internet of things (IoT) with data explosion and bridges theoretical research with evident applications.

Chapter One: Overview of Internet of Things (IoT): This chapter describes the definition of IoT. The term Internet of Things (IoT) or Internet of Everything (IoE) is still ambiguous. There is no single unified definition of what it really is. However, it can be defined by elaborating what it can provide. The Internet of Things is thought to be the next evolution of the Internet as it provides a networking infrastructure allowing for trillions of devices to collect data and communicate with each other to make processed smart decisions.

Chapter Two: Challenges and Applications of (IoT): This chapter discusses some of the major IoT applications that have the potential to bring transformation in our future trends. The rapid development of these applications is expected to face numerous challenges. This chapter brings key challenges into focus and discusses potential barriers that could impede the rapid adoption of IoT.

Chapter Three: Explosion of Data (Big Data): This chapter elaborates on data management, which requires the process of transferring data in an efficient way, upon the user's demand. Therefore, the right data must reach the right user at the right time, in order to be valuable. Data comes from many different sources, i.e. web apps, sensor networks and many others. This data needs to be collected, categorized, stored, and analyzed in order to get an insight of its content and hence present it in an efficient way.

Chapter Four: Boosted Prediction Analysis for Big Data: Prediction techniques represent a useful tool for knowledge discovery in a massive and complex healthcare dataset. In this chapter, a prediction model has been designed and implemented which analyzes medical records of patients and provides information for decision making in health institutes. The proposed model consists of three primary stages, the first being preprocessing data that focuses on preparing the information for the mining process.

Chapter Five: IoT Security: This chapter outlines existing security approaches being used for IoT, together with the weaknesses they inherit. Since the security of IoT communications could be addressed in the context of the communication protocol itself, we focus on existing protocols and mechanisms used to secure communications involved in this vital subject.

Chapter Six: Threat Taxonomy for Cloud of Things: In this chapter we present a comprehensive threat model which is then utilized to create a first-ever threat taxonomy for the Cloud of Things. This taxonomy outlines different security and privacy threats faced by this nascent technology and can be used as the basis for further research on security and privacy in the Cloud of Things.

Chapter Seven: Smart Homes Based On Smart Cities' Design Patterns: This chapter reviews smart cities for the Internet of Things. It discusses maximizing the efficiency of distribution and consumption of energy from one point of view and a vision for smart cities in the future from another point of view. Moreover, it presents the design for a smart home and ends up with a proposed system, as a case study.

Chapter Eight: Social-Sensor Networks: This chapter deals with the integration of Wireless Sensor Networks (WSNs) and Social Networks. Nowadays, WSNs have caused a paradigm shift in our society. They have become a popular mean of communication among people. Many aspects of our lives are significantly related to the WSNs, such as communication, transportation, military, and agriculture.

Chapter Nine: Cellular System Driving IoT: This chapter explores the IoT literature in terms of the communication technology involved: 4G-LTE-A. Additionally, an outline for improving the IoT of future keys in current cellular systems is discussed. This work exhibits how the current 4G LTE-A frameworks can contribute to the design of smart cities. Furthermore, an overview of 4G and 5G is presented. Important features, such as how they will drive IoT to build smart cities and smart objects are considered.

Editors:

Ali Al-Sabbagh[1,2] **and Ruaa Alsabah**[1,3]

[1] Florida Institute of Technology, Melbourne, FL, USA

[2] Ministry of Communication, SCIS-Babylon, Iraq

[3] Karbala University, College of Science, Iraq

December, 2016

Table of Contents

CHAPTER ONE

OVERVIEW OF THE INTERNET OF THINGS (IOT)

Mustafa Kamoona

Department of Electrical and Computer Engineering, Indiana University, Purdue University- Indianapolis, USA

ABSTRACT

The Internet of Things (IoT) or the Internet of Everything (IoE) is a natural evolution of the Internet and is becoming more ubiquitous in our everyday home, business, health, education, and many other aspects. The huge amount of data gathered by the countless Internet of Things devices and processed by Internet of Things networks might be sensitive which calls for feasible and adequate security measures. This chapter gives a brief introduction to the Internet of Things networks and a glimpse about its security. Multiple Internet of Things connectivity solutions are listed, and then a dominant solution, the Wi-Fi, is described in more details. An overview about the Internet of Things security is also explained.

Keywords: Internet of Things (IoT), Computer Networks, Security, Wi-Fi

INTRODUCTION

Characterized by its rapid paced technology development, today's world was not so a couple of decades ago. A substantial technology leap happened when the Internet became public in the 1980s allowing people to surf the web, send emails, and share files. It is always exciting to look back and see how much the world has advanced and how the Internet helped in this process. The Internet was and is still providing a fertile landmark that enables people to communicate in a simple, fast, and convenient way. It is a fact that the Internet continues to evolve shaping our everyday life in the process. Although the Internet of Things networks are now ubiquitous in networking environments, in literature, the term Internet of Things (IoT) or Internet of Everything (IoE) is still ambiguous. There is no single unified definition of what the IoT really is, however, we can define the IoT by elaborating what the IoT can provide. The Internet of Things is thought to be the next evolution of the Internet [1] as it is going to provide a networking infrastructure allowing trillions of devices to collect data and communicate with each other and with other devices to make processed smart decisions. The devices can be any object or anything (thus the name Internet of Things) embedded with the needed hardware and software that are required for processing and networking capabilities. In other words, IoT will be a network of the currently existing rather powerful Internet devices like smart phones, personal computers, and servers with addition of new less complex devices like heart or brain activity monitoring sensors, automobile motion or brake sensors, or any environmental sensors.

From the before mentioned examples, it can be seen that an IoT device does not have to be as complex as the current Internet enabled devices. Thus there is a wider range of devices that can be connected to the IoT networks than that of the Internet. It is predicted that with IoT there will be billions of devices connected and communicating with each other. A typical IoT home environment is shown in Figure (1).

Figure (1) Typical IoT Enabled Home [2]

Whether it is home, business, health, or educational IoT environment, a reason why the IoT environments are getting much attention around the world is due to the fact that a larger scale of integration is possible between the physical objects and the computing systems and thus more intelligent decisions can be made. This everyday life impact of the Internet of Things is possible due to its ability to gather a huge amount of data from devices' surroundings around the globe then analysing and processing this collected data to be able to make a sophisticated decision.

Therefore, the IoT will allow a new era of data exchange and decision making. That is why in 2008, the U.S. National Intelligence Council (NIC) reported that by 2025 Internet nodes may reside in everyday things, food packages, furniture, paper documents, and more. Today's developments point to future opportunities and risks that will arise when people can remotely control, locate, and monitor even the most mundane devices and articles. Popular demand combined with technology advances could drive widespread diffusion of an Internet of Things that could, like the present Internet, contribute invaluably to economic development and military capability [3]. Figure (2) shows some of the many possible application uses of the IoT devices.

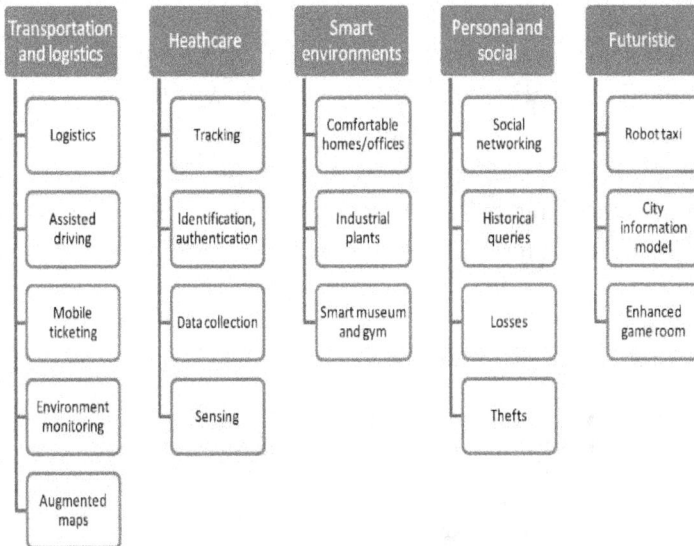

Transportation and logistics	Heathcare	Smart environments	Personal and social	Futuristic
Logistics	Tracking	Comfortable homes/offices	Social networking	Robot taxi
Assisted driving	Identification, authentication	Industrial plants	Historical queries	City information model
Mobile ticketing	Data collection	Smart museum and gym	Losses	Enhanced game room
Environment monitoring	Sensing		Thefts	
Augmented maps				

Figure (2) Possible Current and Future IoT Uses [4]

BRIEF HISTORY:

The Internet of Things might be thought of as the point in time where more things or objects are connected to the Internet than people. An explosive growth of tablets and smart devices happened to increase the number of connected devices from around 500 million connected devices in 2003 while the human population was around 6.3 Billion to 25 Billion connected devices when the population was 7.2 Billion in 2015. According to [1], the point in time when the number of connected devices surpassed the human population was in 2010. Figure (3) shows the timeline of the connected devices versus human population increase. Although the Internet of Things roots can be tracked back to Massachusetts Institute of Technology (MIT) laboratories back in 1999 with the radio frequency identification (RFID) sensing technology [1], whereas the idea of low power communication sensor networks goes back way further in time. The emergence of the distributed low power sensor networks goes back to as early as the year 1967 [4]. Then a series of intermittent events led to the idea of wireless sensor networks (WSN) which in turn led to the concept of smart dust networks.

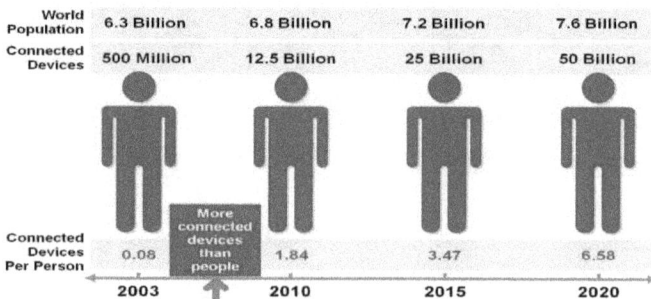

World Population	6.3 Billion	6.8 Billion	7.2 Billion	7.6 Billion
Connected Devices	500 Million	12.5 Billion	25 Billion	50 Billion
Connected Devices Per Person	0.08	More connected devices than people 1.84	3.47	6.58
	2003	2010	2015	2020

Figure (3) Connected Deices Versus Human Population Timeline [1]

Standards started to emerge for such networks in 2003/2004, when firstly the 802.15.4 standard and secondly the ZigBee standard were released. The birth of those standards facilitated the development of the idea of the Internet of Things.

Table (1) shows a timeline of the events that helped in the process of the Internet of Things development. From the majority of those events, we can notice that there are three basic requirements for Internet of Things Networks:

• Low energy communication. As there is little possibility that the IoT devices will be connected to the mains power, their battery life will have to be long for the Internet of Things applications to be practical, as charging or changing the batteries for a huge number of devices would not be a simple process [4].

• Reliable Internetworking enabled communication stack. The IoT devices should be able to communicate with each other and with other devices on other networks that are connected to the Internet. This connection should be highly reliable and bi-directional as data will be traveling both from and into those devices [4].

• Light secure End to End environment. Those networks might communicate sensitive information, so a light secure end to end communication is a necessity for those networks in order to preserve the confidentiality of the data those networks are conveying.

Table (1) Events Helped in IoT Idea Development [4]

Year	Event
1967	REMBASS Remotely Monitored Battlefield Sensor System
1978	Dist. Sensor Networks for Aircraft Detection Lincoln Labs - Lacoss
1992	RAND Workshop - Future Technology Driven Revolutions Military Conflict. Concepts behind Smart Dust emerge.
1993-1994	DARPA ISAT studies - many WSN ideas and applications Discussed. Deborah Estrin leads one of the studies.
1994	LWIM - Low Power Wireless Integrated Microsensors
1997	Smart Dust proposal written, Kris Pister (Berkeley)
1998	Seth Hollar makes wireless mouse collars
1999	Endeavour project proposed by Randy Katz, David Culler (Berkeley) PicoRadio project started by Jan Rabaey (Berkeley)
2000	Crossbow begins selling Berkeley motes
2001	Multiple demos proving viability
2002	Dust, Ember, Millennial, Sensicast founded
2003	IEEE802.15.4-2003 standard Moteiv (now Sentilla) founded
2004	ZigBee 1.0 standard ratified TSMP 1.1 shipping
2005	Arch Rock founded
2006	ZigBee 2006 standard ratified IEEE802.15.4-2006 standard
2007	WirelessHART standard ratified IETF 6LoWPAN's RFC4944 published WirelessHART shown to achieve 99.999% reliability
2008-2009	IETF workgroup Routing Over Low-power Lossy links (ROLL) created. IEEE802.15.4e work group created
2010-2011	IEEE802.15.4e's MAC protocol ratified IEFT 6LoWPAN's RFC4944 updated IETF ROLL's RPL routing protocol ratified

THE FUTURE OF INTERNET OF THINGS

The importance of the Internet of Things and the data the IoT devices communicate is growing exponentially. The procedure of processing the data is shown in Figure (4), it all starts with the gathering of huge amount of data.

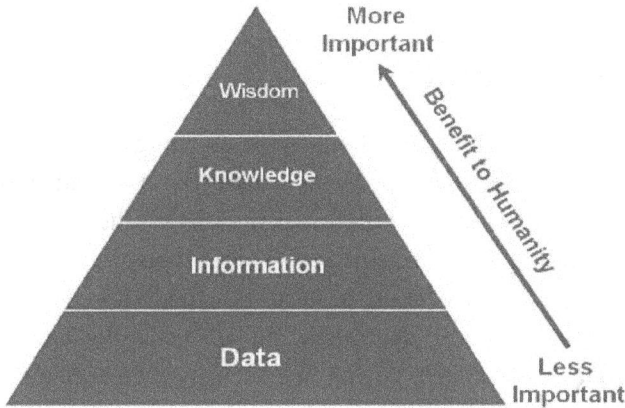

Figure (4). Human Processing of Data [1]

From the bottom up we can see how the data is being processed and becoming more and more important. Data is the raw material that is processed into information. Individual data by itself is not very useful, but volumes of it can identify trends and patterns. This and other sources of information come together to form knowledge. In the simplest sense, knowledge is information of which someone is aware. Wisdom is then born from knowledge plus experience. While knowledge changes over time, wisdom is timeless, and it all begins with the acquisition of data [1].

So the Internet of Things is not limited to the currently existing machine to machine (M2M) architecture that is used for remote control and monitoring, but The IoT creates an intelligent, invisible network fabric that can be sensed, controlled and programmed.

IoT-enabled products employ embedded technology that allows them to communicate, directly or indirectly, with each other or the Internet [2].

CONNECTIVITY OPTIONS

As some might think that the Internet of Things connectivity is a single standard, the reality is, there is a broad range of connectivity (wired and wireless) solutions for the IoT networks. As more and more Internet of Things networks need to be connected on daily basis, a need for flexibility and adaptive configuration arises depending on the complexity, physical environment, available power, and security requirements. In the majority of the cases, the wireless solution is more suitable for the Internet of Things networks than the wired one as it is easier to set up in tricky physical situations and cheaper to install and maintain. However, a care should be taken when choosing the right wireless technology that is adequate for the present circumstances.

Internet of Things Wireless Connectivity

The wide collection of wireless connectivity solutions ranges from the IPv6 Low Power Wireless Personal Area Networks (6LoWPAN) to the Bluetooth Low Energy (BLE) and Bluetooth technology to the ZigBee technology then to the dominant Wi-Fi technology and more. Figure (5) shows the unlicensed frequency bands regions around the world.

Each of the wireless connectivity solutions has its own advantages and disadvantages depending on the range required, circumstances, and environment conditions. Figure (6) shows the different wireless area networks and their respective scopes.

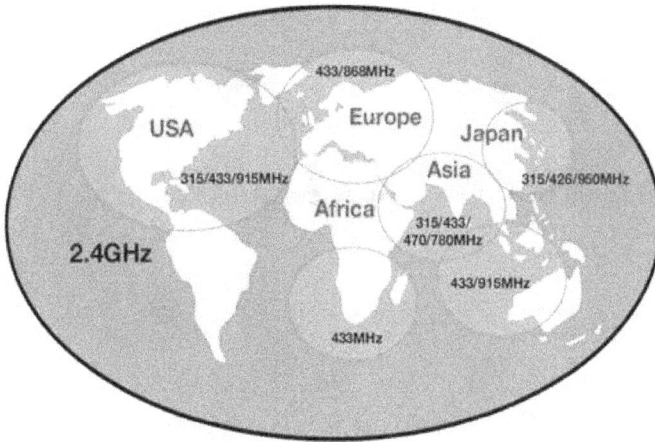

Figure (5). Unlicensed Wireless Frequency Bands [5]

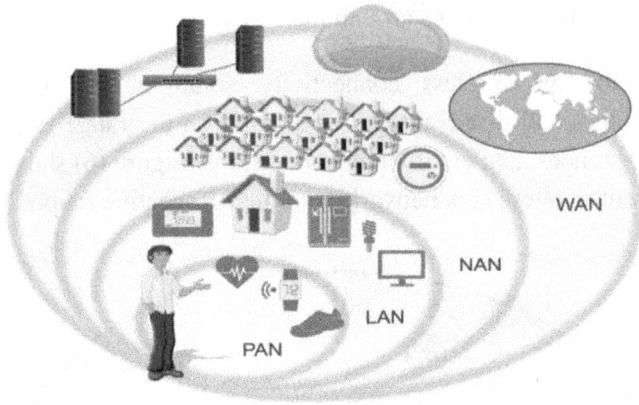

Figure (6). Different Wireless Area Networks [5]

Wi-Fi Solution

The Wi-Fi wireless technology is one of the dominant solutions for the Internet of Things future connectivity. Based on the IEEE 802.11 standard, although the Wi-Fi defines the data link layer of the TCP/IP stack, it is so prevalent that the name Wi-Fi is always associated with TCP/IP wireless networks. The ubiquitousness of the Wi-Fi networks is due to the fact that firstly, almost all of the phones and laptops currently have integrated Wi-Fi modules, and secondly, the interoperability of applications designed by the Wi-Fi alliance.

This success rides on the fact that the Wi-Fi networks are already deployed as part of the infrastructures for homes and business buildings. A natural evolution of the Wi-Fi is to be an integral part of the Internet of Things connectivity [5].

Naturally, the TCP/IP implementation of the Wi-Fi software is complicated and large for the simple design of the Internet of Things and requires much memory and processing. Adding the Wi-Fi solution to the IoT wireless connectivity was not feasible until recently.

Latest silicon advancements made it possible to add Wi-Fi modules to the embedded solutions, the Wi-Fi stack is embedded into the devices and modules to reduce a large amount of the overhead from the micro processing units to allow the smallest micro controlling units to deploy the Wi-Fi connectivity. The increased integration level in those modules, also removes the radio design experience which facilitates the Wi-Fi integration [5].

In addition, in most cases, the IoT devices will need only a small fraction of the Wi-Fi offered bandwidth and data rates, thus with intelligent power management that turn the module on and off (sleep/wake up) to draw small bursts of battery currents, a drastic improvement in the battery life can be achieved. Some current products claim to maintain operation using two AA batteries for more than twelve months [5].

In conclusion, with the prevalence of the Wi-Fi networks, the development of the silicon technology, and smart efficient power management design make the Wi-Fi technology a very promising connectivity solution that helps the advancement of the Internet of Things rapid development [5].

Wi-Fi USE FOR SMART Cities

The technology keeps on changing our life. In the light of IoT we can picture the future cities as smart cities. In a smart city, several physical objects will be interconnected with each other. The inter-connectivity between billions and billions of devices will allow the smart city to integrate and analyze humongous data form these devices.   Thus new services, and exciting solutions for today's problems can emerge. Figure (7) shows how huge data is collected from several technologies, and how this data is integrated and analyzed to provide new exhilarating services.

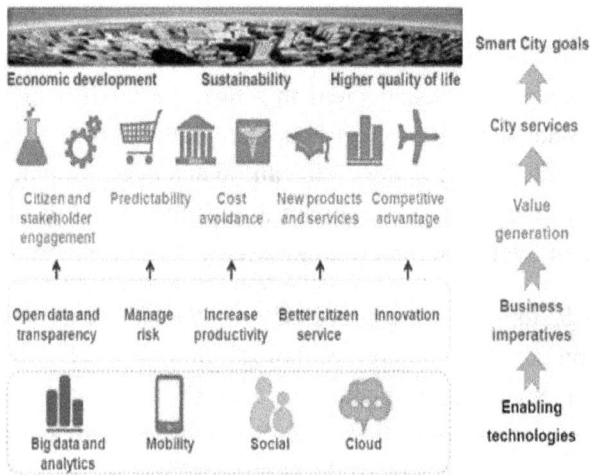

Figure (7). Smart City Architecture and Possible Services [6]

Several examples can be thought of as services offered by smart cities, like health services, educations, traffic rerouting to avoid congestion and accidents, monitoring crime hotbeds to reduce criminal activities, help the city citizens to find parking spots, and help the people to take a more proactive part in contributing with their opinions to the government's officials [6].   As aforementioned, a very suitable infrastructure for the smart cities is the currently implemented Wi-Fi networks. Although the Wi-Fi is ubiquitous, however, there are still some blind spots, areas where there is no Wi-Fi coverage and for that a solution can be accomplished by using umbrella Wi-Fi hotspots to cover these blind spots, hence, ensuring that every spot in the smart city has a Wi-Fi coverage and thus citizens can move freely in the city without the fear of being disconnected.

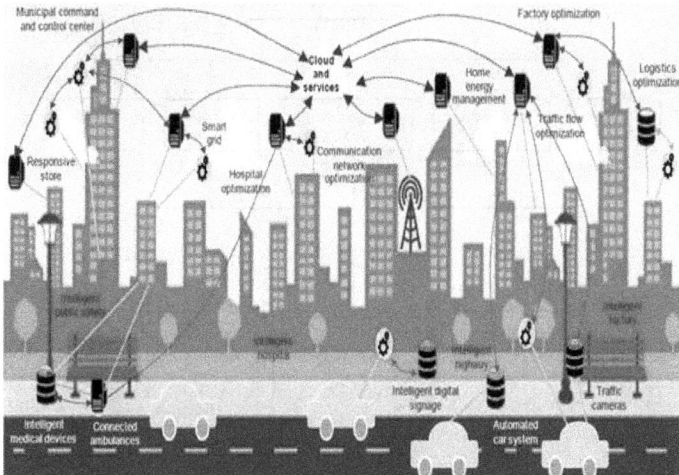

Figure (8). Connectivity in Smart City Example [6]

Introduction to the Internet of Things Security

Computer security, also known as cybersecurity, defines secure network communication as the secure exchange of messages between two entities over an insecure medium or network [7].

Regular computer networks have many security requirements, yet, the Internet of Things networks and because of their intrinsic critical nature mandate even higher security measures. The IoT is an immense network of interconnected networks and those networks usually have devices that are resource constrained thus entail low power computations and low energy consumption. Such networks face numerous kinds of attacks ranging from simple physical attacks to sophisticated cryptanalysis attacks. Figure (9) shows a simple diagram for current known IoT attack types.

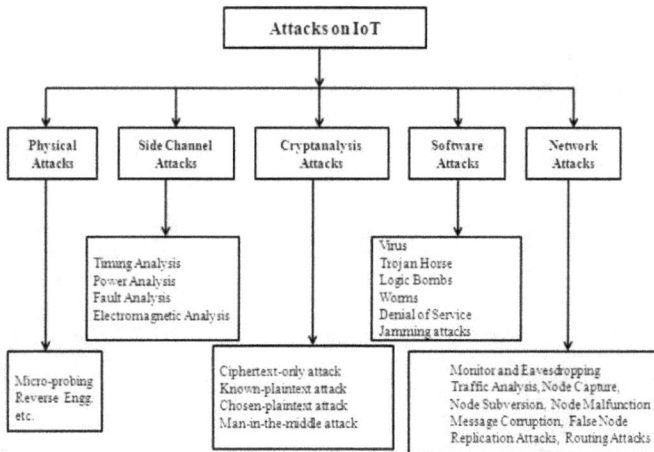

Figure (9). Types of Attacks on IoT Networks [8]

Internet of Things Security Services

Cybersecurity in general has a number of services that the network administrator should keep in mind in order to protect the network from exploited vulnerability. Internet of Things networks should provide the below security pillar services because of the sensitivity of their applications.

Confidentiality the contents of the messages between the two host devices (client and server) should only be read by the authenticated devices. No other inter- mediate adversary should be able to sniff and then read those sensitive contents. Some kind of devices authentication and messages encryption are required.

Integrity the exchanged messages should not be tampered by any intermediate entity with or without purpose. Integrity helps in preventing the man-in-the- middle attacks where a middle device would inject packets into the network masquerading a legit host device. Replay attacks are considered an attack on system integrity since the attacker will record a transaction and then replay that transaction at a later time.

Availability The data that is supposed to be available to authenticated devices should be available to those devices at all times. This service is against denial of service attacks (DoS) where the attacker targets the availability of the provided services to the authentic users.

The above services are provided by different devices and layers in the network with the aid of symmetric key cryptography, public key cryptography, and hash functions that are briefly explained in the following section.

Introduction to Cryptography

Cryptographic techniques have gone hand in hand with secure message communication for a long time even before networking and computers were devised. A long description of cryptographic history and evolution is explained in details in [7]

[9]. With the emergence of computer networking, cryptography has become inseparable with Cybersecurity services like authentication, encryption, and integrity [10].

Cryptography enables a user to encrypt his message which is called a cleartext or plaintext and send it as an encrypted message which is labeled as ciphertext, those ciphertexts should be unfathomable to any intruder that intercepts those messages.

Hash functions are inextricable from cryptography. A hash function in general is a function that generates a fixed length output string for any given input, a cryptographic hash function, however, has more restrictions including that it should be infeasible to find two different inputs that result in the same output string.

Many standardized encryption algorithms and techniques are available in reference for comments like [RC 3447] and [RFC 1321] [7]. In this thesis, as in many cryptography sources, the two communicating entities are addressed as Alice and Bob and the intruder is labelled as Trudy. The algorithm that encrypts the messages, that is, changes a plaintext into a ciphertext, is referenced as a cipher. Some of those terms are illustrated in Figure (10). Usually the cipher is public knowledge but the secret is the keys Ka and Kb which can be the same in both sides in the case of symmetric key cryptography or different in the case of public key cryptography.

A simple cipher might encrypt a message (m) by using its key $KA(m)$ and on the receiving side the cipher can decrypt the ciphertext with its secret key KB to recover the original message $(KB(KA(m)) = m)$ [7]. The cipher might be a simple bitwise XOR on both sides.

Symmetric Key Cryptography

In symmetric key cryptographic systems, both the communicating entities (Al- ice and Bob) use the same shared key (symmetric key) to encrypt and decrypt the messages. Where it is assumed that at some point before the communication starts, the two entities have agreed on a shared secret key through a secure communication channel. Symmetric key cryptographic techniques are usually computationally faster than public key ones.

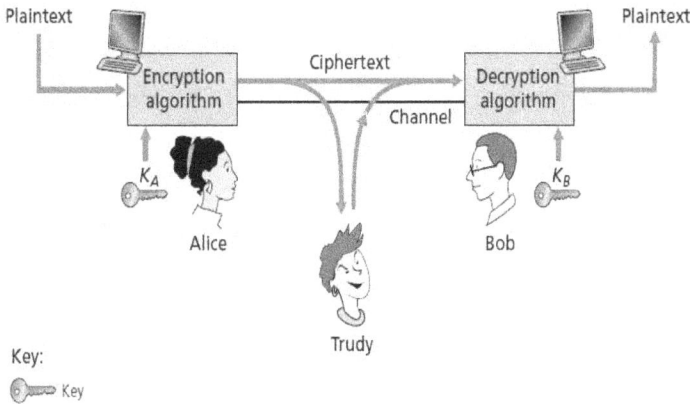

Figure (10). Cryptography Terminology [7]

Public Key (Asymmetric) Cryptography

In public key cryptographic systems, a pair of keys which are mathematically related are used. A public key (K+) which is available to the public and a private key (K-) which is a secret to everyone but to the entity itself. In order to exchange messages, a sender (Alice) should encrypt a message with the receiver's

(Bob's) known public key (KB+) and send the ciphertext (KB+(m)) to Bob, then Bob, using his own private key (KB-), should decrypt the ciphertext (KB-   (KB+(m)=m). It is interesting to note that if a message is encrypted by a public key (KB+) then it can only be decrypted by the same entities private key (KB-) and vice versa, thus the same key cannot be used for encryption and decryption. Many techniques target public key cryptography, and probably the most prevailed one is known as the RSA algorithm. Figure (11) shows a public key cryptography scenario.

In the cybersecurity community, there is a wide diversity in the Internet of Things state-of-the-art works. Whereas all the works agree in terms that the Internet of Things consisting mostly of

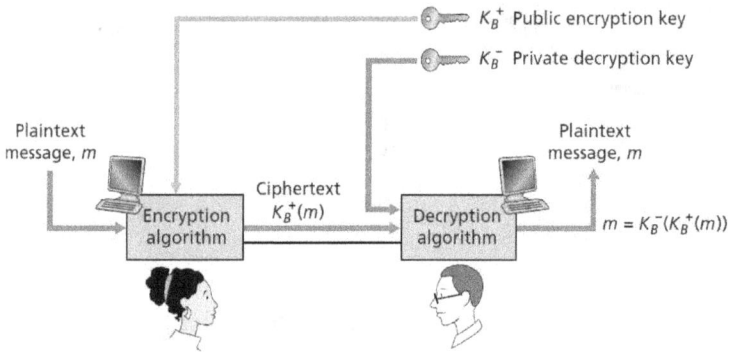

Figure (11). Public Key Cryptography Scenario [7]

embedded systems that generate sensed data- should have the messages carrying them encrypted before being sent, some of the work targets end to end encryption like the Secure Internet of Things Project (SITP) where the data is encrypted in the end device (sensor) and stays encrypted until it reaches the other side of the internet (server) where it gets decrypted [11], another work proposes and compares new light weight solutions for the IoT including link layer security, IP layer security (IPsec), and UDP datagram transport layer security (DTLS) and gives the pros and cons of each of them [12]. On the other hand, another work focuses on a specific part of the IoT solution that suggests a new wide area concept within the operator's Long Term Evolution (LTE) macrocelullar network uplink for intermittent IoT traffic that works simultaneously with the normal LTE traffic in a way that do not greatly affect the overall efficiency of the system [13].

REFERENCES

[1] D. Evans, "The internet of things: How the next evolution of the internet is changing everything," CISCO white paper, vol. 1, pp. 1–11, 2011.David Easley and J Kleinberg. Strong and weak ties. Networks, Crowds, and Markets: Reasoning about a Highly Connected World, pages 47–84, 2010

[2] J. Chase, "The evolution of the internet of things," Texas Instruments, 2013.

[3] C. T. Fingar, Global trends 2025: A transformed world DIANE Publishing, 2009.

[4] M. R. Palattella, N. Accettura, X. Vilajosana, T. Watteyne, L. A. Grieco, G. Boggia, and M. Dohler, "Standardized protocol stack for the internet of (important) things," Communications Surveys & Tutorials, IEEE, vol. 15, no. 3, pp. 1389–1406, 2013.

[5] G. Reiter, "Wireless connectivity for the inte-rnet of things," Europe, vol. 433,2014.

[6] R. Y. Clarke, "Smart cities and the internet of everything: The foundation for delivering next-generation citizen services," Alexandria, VA, Tech. Rep, 2013.

[7] J. F. Kurose and K. W. Ross, Computer networking: a top-down approach.Addison-Wesley, 2007.

[8] S. D. Babar, Security Framework and Jamming Detection for Internet of Things.
Videnbasen for Aalborg UniversitetVBN, Aalborg UniversitetAalborg Univer-
sity, Det Teknisk-Naturvidenskabelige FakultetThe Faculty of Engineering and
Science, 2015.

[9] S. Singh, The code book: the science of secrecy from ancient Egypt to quantum cryptography. Anchor, 2011.

[10] D. Kahn, The codebreakers. Weidenfeld and Nicolson, 1974.

[11] P. Levis, "Secure internet of things project (sitp)," 2015. Online: http://iot.stanford.edu, accessed 11-April-2016.

[12] S. Raza, Lightweight Security Solutions for the Internet of Things. PhD thesis, M□alardalen University, Vasteras, Sweden, 2013.

[13] C. S. Bontu, S. Periyalwar, and M. Pecen, "Wireless wide-area networks for internet of things: An air interface protocol for iot and a simultaneous access channel for uplink iot communication," Vehicular Technology Magazine, IEEE, vol. 9, no. 1, pp. 54–63, 2014.

CHAPTER TWO

APPLICATIONS OF IOT AND FUTURE CHALLENGES

Naveed Nawaz

School of Electronic and Electrical Engineering, University of Leeds, UK & Department of Electrical Engineering, UET Lahore, Pakistan

ABSTRACT

Internet of Things (IoT) will revolutionize the present digital world and has the potential to bring a paradigm shift in the future business trends. The focus of big industry players will be to develop whole integrated solutions for future applications. It is expected that there will be an exponetial growth in the number of internet connected devices. This cahpter discusses some of the major IoT applications that have the potential to bring transformation in our future development trends in industry. The rapid development of these applications is expected to face numerous challenges. This chapter brings key challenges into focus and discusses potential barriers that could impede rapid adoption of IoT.

Keywords: Internet of Things (IoT), Smart Cities, e-Health industry, Energy, Efficiency, Security, Privacy, Latency.

INTRODUCTION

Technology has evolved tremendously in the past few decades. It has brought significant transformation to our lifestyles. The ways of communication using smart phones have taken the place of traditional conversations. The use of internet through smartphones, tablets and other electronic gadgets have significantly changed the way we socialize, read and share information. The improvements in data transmission rate and miniaturization of electronic devices have played an important role in the evolution of technology. The increasing trend to manufacture smart, energy-efficient sensors and then connecting them to the internet has opened up new avenues in today's world [11]. It has now been envisaged that almost all objects of daily usage will be connected to internet and will have some computational intelligence in future. Such a revolution of having connected things all around us will realize the concept envisioned by Kevin Ashton—the founder of the term IoT. In the next section, we discuss some significant applications of IoT.

APPLICATIONS OF IoT

It is expected that there will be an exponential growth in the number of internet connected devices and will exceed beyond 50 billion by the end of 2020 [6]. IoT will revolutionize the present digital world and will bring a paradigm shift in the future business trends. The focus of big industry players is to develop complete and integrated solutions for various applications. Some of the IoT applications that have the potential to bring a significant transformation in our future trends are discussed in the subsequent sections:

Electronic Health Industry

With the advent of communication technologies and development of energy efficient, low-cost and small-sized wearables and body sensors, various health care applications are expected to bring about a paradigm shift in the whole e-health industry. In a health care system, different body sensors and wearables on a patient can be connected to the internet. The data generated in this way can be collected at a central database for analysis and diagnosis of a disease. For example, a heart patient may be asked to wear a sensor belt that can continuously monitor his ECG and respiratory signals. Similarly, a psychiatric patient may wear different body sensors through which various environmental parameters can be collected persistently. The connectivity of such a body sensor network with a central database enables physicians to better diagnose any disease. Moreover, this obviates the need of physical presence of a patient in a hospital for his routine check-ups. All this will eventually result in better quality of life of the people.

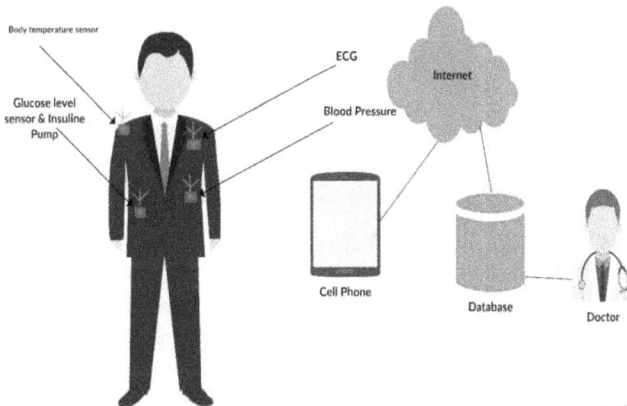

Figure (1) Electronic Health Care System

Many solutions are being proposed to develop different health care systems. In order to develop a holistic system, it is really important to use open standards for e-Health. Interoperability ensured in these standard specifications will allow heterogeneous devices and services to interact with each another without any operational barriers. For example, a complete telemedicine system consists of Wireless Body Area Network (WBAN) where sensors send health relevant information to medical servers through internet. The hospital staff such as physician and nurses could consult this database from anywhere and at any time. An e-Health system needs to ensure interoperability so that different network devices could cooperate and communicate with each other regardless of manufacturers. It needs to have bounded delays and reliability so that delay-intolerant emergency applications could be served. Moreover, issues of privacy, integrity and authentication of sensitive need to be dealt efficiently, especially in a resource-constrained network where devices have limited hardware capabilities and meagre power supply. IoT should provide an efficient framework to cope with the requirements of e-Health system.

Smart City

Inhabitants of an area can be provided better facilities by developing efficient traffic management system, waste management system, water supply system, law-enforcement system and smart parking system. Big Data analytics and the development of various IoT applications can help to improve the overall infrastructure and transform an area into a "smart city" [12], [2]. For example, the traffic management system can monitor traffic in real time by use of different sensors. It can observe the occurrence of unlikely events such as accidents or traffic jams. These sensors can be connected with the internet so that they could pass their observed information to a central database. After analysis of a real-time data, information about traffic congestion can be broadcast to road users. The highways can display variable warning messages. All this can help people to adopt alternate available routes and thus save their time and energy.

Such an intelligent transport system is an important application of IoT that can efficiently resolve issue arising from increased vehicle density [1]. Similarly, deployment of various IoT devices and sensor in a city can provide better law enforcement and improve overall public safety. It can enable to provide quick response in case of certain emergencies.

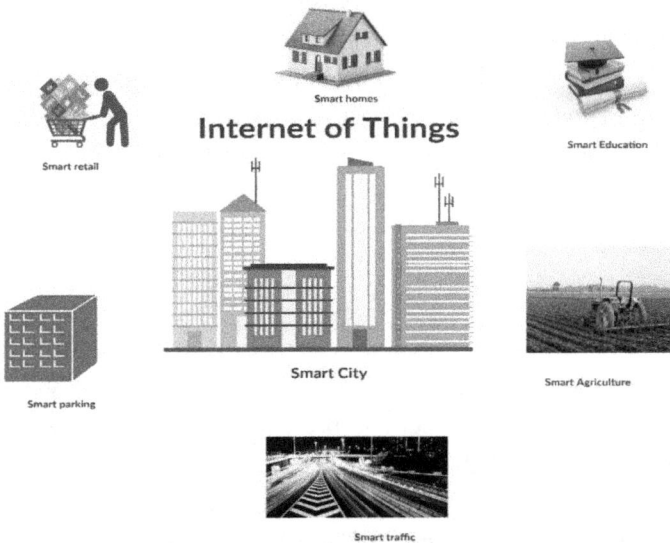

Figure (2) Smart City

Smart Manufacturing

Industrial automation and smart manufacturing is one of the important application of IoT. The manufacturing process in an industry can be optimized by deployment of various smart and connected sensors. The interaction among industrial equipment can be enhanced by deploying various features of IoT. The machines having in-built intelligence can perform automated decision-making and enable smart manufacturing. The industrial units can have sensors to collect information about possible wear and tear and thus provide an opportunity to perform proactive maintenance.

Smart Mobility

Autonomous driving by deploying sensors on cars and connecting them to the internet is another possible application of IoT. It can provide smart mobility to different road users. Transport industry can track current location of their fleets and check the condition of goods inside shipped containers by use of sensors.

Smart Homes

IoT provides an opportunity to connect almost every home appliance with the internet and presents many ways to remotely access and control these appliances [7]. Nowadays, it is quite possible to remotely control domestic lighting, adjust room temperature, check security of home, start a washing machine, unlock or start the car. Smart home applications will provide a possiblity to manage all these application from anywhere in the world. IoT will provide an opportunity to integrate all these applications and will provide ways to link intelligence of all the sensors in various appliances in a single application.

Figure (3) Smart Home

Smart Grid

The world is focusing on conservation and efficient management of energy. The utility providers are managing the distribution and automation of smart grid. One of the key requirements of an automated distribution is to locate and monitor the substation dynamically from main control centre. IoT will be instrumental in providing cost-effective ways for connectivity of smart grids [9]. It would pave ways to repair critical infrastructure quickly by informing maintenance teams proactively. The grid will eventually transform to a network of fully automated substation with connectivity to the internet. Moreover, the utility companies will use smart home meters to manage billing of their customers. They may keep track of trends of energy demands. They may also inform the consumers about their billing records and suggest them the ways to reduce their energy bills.

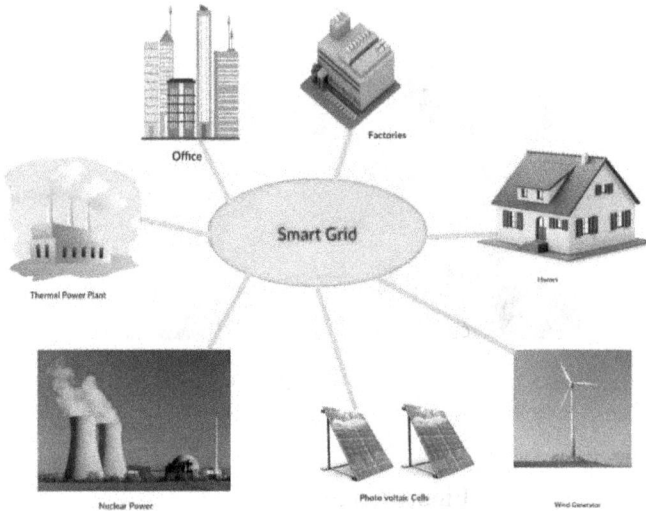

Figure (4) Smart Grid

FUTURE CHALLENGES OF IoT

The tremendous technological development has paved the way towards exponential growth of IoT. Although the connectivity of billions of devices is on the cards, there is still a long way to go. We present various critical challenges in smooth deployment of IoT devices that must be addressed with proper attention.

Security

In IoT, it may not be practical to manually connect or configure billions of devices and sensors. Thus it is necessary that such a huge number of devices should have the capability of self-configuration. They are likely to be connected to the internet automatically and start communicating with other objects without any robust authentication mechanism.

Moreover, it is expected that there will be numerous IoT applications in future. It is quite challenging to keep an eye on the counterfeit devices while communicating with the diverse IoT applications. A hacker may connect many fake devices to the network [4]. This makes an IoT network vulnerable to cyber-attacks and raises an eyebrow about its overall security in the long run.

In some critical IoT applications, the breach of security can have grave consequences. For example, hacking the control system of an industry can put the entire manufacturing process to a halt; muddling up with the medical electronic devices such as pacemaker or Implantable Cardioverter Defibrillator (ICD) can endanger a human life.

While some of the applications can embed an additional hardware and implement robust encryption techniques or authentication algorithms, most of the devices cannot afford to take such stringent security measures as they are likely to have limited resources and can barely execute their own tasks.

Therefore, the security of IoT devices against cyber-attacks is one of the important challenges for the future IoT applications [8]. It needs a proper focus especially when the devices are small, inexpensive and have limited computational power.

Privacy

The protection of data and privacy is one of the fundamental rights of people. The younger generation may not bother much about a privacy infringement but a certain backlash is indispensable. The consumers will adopt products only if they are sure that their privacy will not be compromised. As discussed earlier, the concept of autonomous connectivity with billions of end points makes IoT vulnerable to cyber-attacks [3]. The data streams of various IoT applications are exposed to theft and may be used by hackers to collect sensitive information. The privacy issues may not come to the fore at the beginning, but they need to be addressed at the very outset for sustainable growth.

Interoperability

The IoT deployment will provide numerous applications in future. The devices for these applications will have different requirements for efficient operation. There are many competing communication standards to efficiently manage these devices such as Radio Frequency Identification (RFID), Near Field Communication (NFC), Low Energy Bluetooth, Zigbee, Ultra Wide Band (UWB) and IEEE 802.11 ah [3].

The service providers may choose one of the candidate solutions for implementation of an application. Some of these manufacturers may adopt their own proprietary solutions as well.

It will be a significant challenge to integrate all these applications in a fragmented market where manufacturers can choose among various available standards. The development of one whole solution is possible only when various implementations are interoperable. The solutions providers must be aware of the future problems that may arise due to lack of focus towards interoperability and should stick to open standards only. The developers of these standard specifications must also focus to fulfil the needs of multi-vendor business environment and provide best practices for implementation of various IoT applications.

Energy Efficiency

The deployment of billions of IoT devices raise concerns about their energy requirements. For a cost effective solution, all the devices should consume energy efficiently [10]. The solution to install power cables for a massive deployment in IoT may not be feasible. Even battery powered sensors would cause huge maintenance expenses. It has been a focus of active research to develop efficient energy scavenging techniques for these devices. One of the suggested technique uses solar cells to convert heat into electricity. The devices may also use thermoelectric generators to use light as a source of energy. Although the energy produced by the energy harvesting techniques is in small proportion, usually of the order of milliwatts/cm^2, however they can prove to be enough for supply power to IoT devices. This is possible by efficiently utilization of energy.

If IoT applications are to come forward in future as low cost alternatives, low cost and energy efficient solutions are the need of the hour. Many networking protocols focus to reduce energy consumption by keeping the devices in sleep mode [10]. In this regard, the development of efficient power saving features is indispensable and will boost the rapid deployment of IoT devices.

Throughput

Large number of IoT devices will share the same channel for data transmission. They are likely to compete for the channel access. With increase in number of contending nodes, the numbers of collisions and retransmissions can become sufficiently large, especially under large traffic volumes. This may cause reduction in overall throughput of network. Moreover, payload information being sent by most of IoT applications is likely to comprise of short data packet. The header information of small data packets may reduce the overall network throughput. It becomes a challenge for IoT network to efficiently manage such a large number of nodes without degradation of overall performance.

Low Latency

IoT network will be supporting large number of devices in future. The nodes will contend for the access of shared medium. They may face longer delays while transmitting their data, especially under heavy load conditions. The newly generated data packets may have to queue up in buffer of a node and have to wait for prolonged duration. Most of these nodes are expected to have limited memory capacity and will not store large amount of data. Such prolonged delays will keep the devices active for relatively larger amount of time and may result in increased energy consumption. Moreover, some of IoT applications such as wildfire detection may require quick response and cannot afford to have longer delays. Hence, latency becomes another significant challenge that needs to be focused for future IoT growth.

Traditional Inertia

Some of the companies may remain hesitant to take initiative to opt IoT solutions. Some of them may have different budget priorities and many remain away from due to their traditional inertia. This can result in another challenge for future growth of IoT.

CONCLUSION

IoT will bring revolutionary changes to the ways of communication. It will connect all the objects, people and their surroundings. Things will communicate with each other without any human interruption. It will provide new avenues for future infrastructure development. There will be numerous IoT applications in different sectors. The future business models will also change significantly. The solution providers will now focus to come up with a complete solution by integration of various useful applications. The growth of IoT devices will be enormous and we may see deployment of billions of devices in the coming years. Most of these devices are likely to have limited resources such as memory, processing capability and power supply which may pose a serious challenge on the future growth of IoT network. There is a dire need to focus on the issues such as security, privacy, interoperability and energy efficiency in this regard. A lot of research effort is require to find optimal solutions to address these challenges.

REFERENCES

[1] Ali Abbas Al-Sabbagh, Ahmed Yaseen Mjhool and Ruaa A. Saeed Alsabah. Smart parking techniques based on internet of things. Journal of Networks and Telecommunication Systems, 1: 1-10.

[2] Ruaa. A. Saeed, Alsabah Ali Abbas Al-Sabbagh and Ahmed Yaseen Mjhool. An extensive review: Internet of things is speeding up the necessity for 5g. Int. Journal of Engineering Research and Applications 5, 2015

[3] Luigi Atzori, Antonio Iera, and Giacomo Morabito. The internet of things: A survey. Computer networks, 54(15): 2787-2805, 2010.

[4] Luigi Atzori, Antonio Iera, and Giacomo Morabito. From "smart objects" to "social objects": The next evolutionary step of the internet of things. Communications Magazine, IEEE, 52(1): 97-105, 2014.

[5] Ericsson TU Dresden and NSN Vodafone. A choice of future M2M access technologies for mobile network operator.

[6] LM Ericsson. More than 50 billion connected devices. Ericsson White Paper, 2011.

[7] Jayavardhana Gubbi, Rajkumar Buyya, Slaven Marusic, and Marimuthu Palaniswami. Internet of things (iot): A vision, architectural elements, and future directions. Future Generation Computer Systems, 29(7):1645-1660, 2013.

[8] Sungmin Hong, Daeyoung Kim, Minkeun Ha, Sungho Bae, Sang Jun Park, Wooyoung Jung, and Jae-Eon Kim. Snail: an ip-based wireless sensor network approach to the internet of things. Wireless Communications, IEEE, 17(6):34-42, 2010.

[9] Stamatis Karnouskos. The cooperative internet of things enabled smart grid. In Proceedings of the 14th IEEE

international symposium on consumer electronics (ISCE2010), pages 07-10, June, 2010.

[10] Jia-Ming Liang, Jen-Jee Chen, Hung-Hsin Cheng, and Yu-Chee Tseng. An energy efficient sleep scheduling with qos consideration in 3gpp lte-advanced networks for internet of things. Emerging and Selected Topics in Circuits and Systems, IEEE Journal on, 3(1): 13-22, 2013.

[11] Charith Perera, Arkady Zaslavsky, Peter Christen, and Dimitrios Georgakopoulos. Context aware computing for the internet of things: A survey. Communications Surveys & Tutorials, IEEE, 16(1): 414-452, 2014.

[12] Andrea Zanella, Nicola Bui, Angelo Castellani, Lorenzo Vangeslista, and Michele Zorzi. Internet of things for smart cities. Internet of Things Journal, IEEE, 1(1): 22-32, 2014.

CHAPTER THREE

EXPLOSION OF DATA (BIGDATA)

Raed Sahib

Department of Electronic and Computer Engineering Brunel University London, UK

ABSTRACT

The Internet of Things (IoT) assumes that everything is connected and therefore an enormous amount of data (Big Data) can be transported. There is a need to address this data in terms of storage, processing, and analyses in addition to elaborate on transmission methods. There is a huge amount of useful information hidden within this Big Data which needs to be mined in order to acquire the knowledge to create new opportunities and overcome rising challenges. Big Data represents a challenge to the current generation of mobile network. The transmission methods are not yet ready to handle such amount of data. With the upcoming 5G mobile network, Big Data transmission will be addressed and solutions for Big Data transmission are required as 5G will be the base infrastructure for IoT. This chapter will give an introduction to Big Data and the combined challenges.

Keywords: Internet of Things (IoT), Big Data, Content Delivery Network (CDN), 5G.

INTRODUCTION

Today's world is referred to as the "information technology age or era", and of course it is the digital portion of information that we are talking about (rather than analogue). This allows for vast new opportunities and support for many additional new technologies that rely on discovering and utilising information such as biotechnology and nanotechnology, in addition to planning and testing final products (even without prototypes) in materials.

How did we reach this era and why did information become so important? The answer to this question begins with understanding the early development of computers and the revolution of computer and communication networks.

In the past century, computers have had a lengthy history and have passed through important stages from the 50's to 70's. Nevertheless, they were of limited use. In the early years, computers were only used by scientists and engineers for research and development purposes. Later on, computers were used by businesses and governmental organisations.

In the 1980's, computers started to become a part of people's daily lives and were used by professionals, students and at homes. They grew to become personal and household items, much like the TV.

With this spread of personal computers (PC's), and the development of computer networks, metropolitan area networks, and the Internet, our world (people and organisations) became more connected and paved the way to share information through networks and business processes.

In addition to this spread and network development, the revolution in the mobile networks (computers and phones) that started in the late 1990's till current days, had a vital impact on information technology and encouraged the integration of devices (computers, telephones, cameras, televisions, etc.). Telecommunication is the second important factor in the information technology revolution due to its primary role in facilitating the transfer of data, information, images, pictures, etc. Therefore, mobile networks are utilized, beginning at the second generation (2G), which represents the first digital mobile system as well as the first mobile system to introduce the use of data by using the well-known term of SMS (Short Message Service). Passing through the subsequent generations, higher data rates are provided. Many applications and use cases are presented as a need, or as industries evolve. These applications and use cases are presented or will be presented as an advantage of a new network's existing capabilities. The prospective next mobile generation (5G) is intended to be a revolutionary system that integrates many different networks and aims to provide much higher data rates than ever before. The reason behind this revolutionary system is to enable the processing of a massive amount of data generated and transmitted due to the widespread use of these applications. In particular, 5G will be the foundation of the Internet of things (IoT). And with IoT, the amount of data exchanged is expected to increase in an exponential way. It can be clearly seen that 5G explains IoT and IoT explains Big Data.

IoT represents the inevitable result of the maturity of both computer systems and communication systems. The convergence of computer technology and telecommunication technology is one of the main reasons that the current Information era was achieved. Computers are used to process data! Communication lines are used to transfer processed data from one computer to another (between devices), or between websites (servers) and users (devices). Computers and telecommunication have merged together and this is evident when looking at how business undertakings have developed, e.g. e-government, e-learning, e-commerce, etc.

CHALLENGE OR OPPORTUNITY:

Data is everywhere, people and devices continuously generate data. These could differ depending on their trends and methods. Nevertheless, they behave as data generators in all the cases. This data varies in size; it could be small or big depending on the situation and the generator. If we consider the rapid increase in the use of devices, especially mobile devices that are connected to different networks for different purposes, and considering the growth in the number of people attached to the internet, we will have an idea about the size of data being generated and transferred every moment; and of course it is an enormous amount.

Let's take an example of a well-known website to all of us, YouTube. YouTube has over a billion users, which is estimated to be a third of the internet population. Users watch hundreds of millions of hours every day and generate billions of views [source: YouTube website]. Growth in watch time on YouTube has accelerated and is up by at least 50% year after year for three years straight [source: YouTube website]. The sharing and uploading of videos grew in a very fast way. In 2007, the total uploaded videos on YouTube were equal to six hours per minute. Meanwhile, in 2010 the rate was 24 hours, then 35, then 48, and to this moment it is estimated to be 60 hours of videos per minute. In other words, it is one hour of uploaded videos for every second [source: YouTube website]. This is a big number, and it represents a challenge. However, it is potentially an opportunity as well.

It is a challenge dealing with this huge amount of data in terms of storing and managing, as well as ensuring the network sustainability is not affected. However, it represents an opportunity for using a customer's data (through approved cookies) to target the customer needs and requirements based on their interests. One might think that this data is of no use, such as a person sharing his location on social media with family and friends. The fact is, this data is important to someone else. For example, there are companies that collect information on our shopping patterns, medical needs. Furthermore, doctors and insurance companies gather medical test results, and governments compile logs of our phone calls and emails. However, simply having lots of data is not the same as understanding it. A good understanding of the customer behaviour will lead to a better introduction of new services; more accurately, analyses will lead to more confident and effective judgment and decision making.

Better decisions could result in cost reduction and reduced risk; it will also affect how quickly it can deliver new applications to market, support a business's growth and improve customer experience. The way an organisation deals with this data has a direct impact on its success. Hence such data could be useful if it is tagged and analysed.

"By 2020, it is estimated the digital universe will reach 44 zettabytes of data"

[https://www.emc.com/en-us/big-data/index.htm]

According to Executive Summary, in April 2014 by IDC about EMC Digital Universe:

[http://www.emc.com/leadership/digital-universe/2014iview/executive-summary.htm]

- In 2013, only 22% of the information in the digital universe will be a candidate for analysis, i.e., useful if it were tagged (more often than not, we know little about the data, unless it is somehow characterized or tagged – a practice that results in metadata); less than 5% of that was actually analysed.

- By 2020, the useful percentage could grow to more than 35%.

- In 2013, the available storage capacity held 33% of the digital universe. By 2020, it will be able to store less than 15%

A familiar example of data collection and analysis is the Google privacy policy that pops up on our browsers from time to time

Figure (1). Percentage of data candidate for analysis

when we use their web services. Google processes the data we use, such as when using Google maps to find a location or when using Google translate or search engine. They collect this information and also information about devices' ID's, IP addresses and locations. The purpose behind this is to have an insight of the users' activities which will then be analysed and used for the benefit of the company and the user. This is demonstrated through the delivery of targeted adverts based on the consumers' interest for example and also to improve the quality of service and security methods provided by Google. This is based on the data collected from their users based on their internet behaviours and usage.

Dealing with big data is much like climbing a mountain, very difficult! At the beginning, one may find everything similar and will not know exactly where to go and where to begin, but the picture will get clearer, and many facts and information patterns will be revealed as steps are taken to the top. It is believed that monks lived in temples built at the top of mountains, in a similar way, ancient civilizations built their temples at the top of high pyramids and crowning ceremonies of a new kings and queens took place all the way at the top of these temples, simply because wisdom was rightfully considered at the top of everything.

From these examples, it can be concluded that it is not the data itself that is of importance but what can be extracted from the data and how to utilise outputs, which is known as information. We live in the "information technology era". This information naturally leads to knowledge, and in turn will lead to wisdom; which is the abstraction of this knowledge.

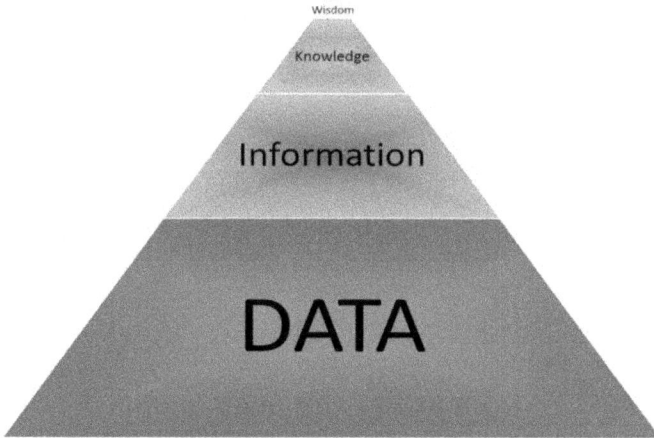

Figure (2). Data Mountain

WHAT IS BIG DATA?

Regardless of the fact that gathering, storing, and analysing data for different purposes is an old process, it could be hard to find a specific definition for Big Data as this term is relatively new. Simply, the term "Big Data" refers to the huge amount of data that is increasing exponentially and rapidly. This term "Big Data" gained momentum in the early 2000s when industry analyst Doug Laney articulated the now-mainstream definition of Big Data as the three Vs [www.sas.com], volume, variety, and velocity which represent the three main characteristics of any data in order to be considered big.

This 3V's – volume, velocity, and variety – are the three main

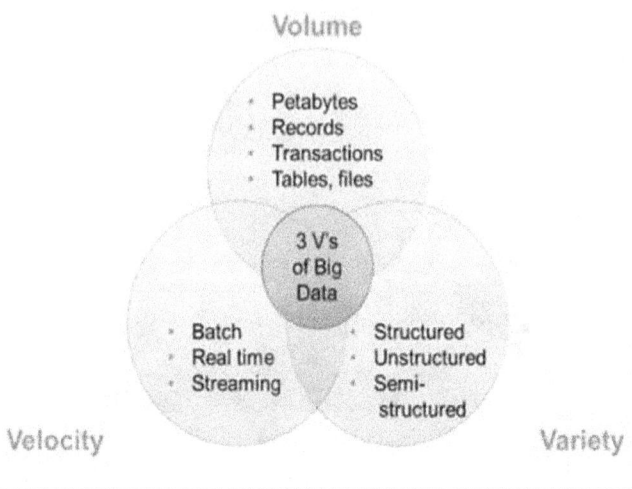

Figure (3). The 3 Vs of Big Data [1]

characteristics that define big data, but sometimes more V's are suggested like veracity and value, and some add complexity as well when referring to the challenges of Big Data [2], [3]

The definition of big data from Gartner stated that: [4]

"Big data is high-volume, high-velocity and/or high-variety information assets that demand cost-effective, innovative forms of information processing that enable enhanced insight, decision making, and process automation."

DEALING WITH BIG DATA

Data management requires data to be transferred as and when

needed by the users in an efficient way so that the right data reaches the right user/s at the right time in order for it to be valuable.

Data comes from many different sources, i.e. web applications, mobile applications, sensor networks and various other sources. This data is in a raw form which will need to be collated, categorised, stored, and analysed. This process will provide insight into the data content and will allow it to be presented in an efficient and user-friendly way to the end user. There are many architectures available for planning and/or mapping the big data, all of which can be summarised by the below structure (Figure 4). An important element of this structure is the data delivery and transportation which may be required at different stages of the architecture.

Figure (4). Big Data Architecture

The process of handling Big Data varies between different business models, depending on the field of interest. Companies are collecting data generated by machines and people, not necessarily related to their main business but mostly related to products or user behaviour.

Big Data combines a mixture of structured, semi-structured, and unstructured data, which is typically beyond the ability of traditional data warehousing and processes' software. There are many businesses that deal with the storing, processing and the analytics of data, such as EMC, IBM, and SAS. On the other hand, others deal with the transfer of this data or its content, such as AKAMAI.

Transportation is an important part in this field. Taking into consideration the current research in the field of IoT and 5G mobile networks, for many applications data is required to be processed and transmitted in a timely manner. 5G requirements consider the term of zero latency – as there is an increased demand for the higher data transfer rate with low latency for many cases ranging from entertainment applications passing through the medical and environmental applications that are in direct or indirect touch to the people's lives. In addition to the latter, there are new terms, such as wwww (worldwide wireless web), intelligent or smart society, virtual reality, and M2M or MTC. Many of the MTC applications such as car self-driving, telemedicine, and online gaming will require almost zero latency and very high bandwidth. MTC applications will also require 100% network availability, and for many of the cases, a real-time decision making, decision automation and even predictive manner based on the user behaviour could be required.

An example of such an application is telemedicine which will benefit from the use of M2M or MTC technology to make it possible for surgeries to take place using machines while the surgeons are located in different locations than the patient/s'. Surgeons can, for example, operate in battlefields when there is an urgent case which is not possible to be transported or could potentially risk the patient's life.

"Big Data sounds sexy, but it's actually very dirty"

Tapan Bhatt, senior director of industry solutions, Splunk Inc.

Big Data represent a big opportunity; however, we need to deal with its 'big' problems. The best way to deal with any problem is to divide it into stages, i.e. as in the Big Data architecture stages, Figure (5) above describes it.. For each stage of the Big Data architecture, there are problems and solutions. Many techniques have been created in order to deal with the big problems of data, and many platforms have been produced for this purpose such as Hadoop, MongoDB, and Cassandra. Most of those platforms deal with data management and storing. A very important factor within the big data problems is communication. Big Data needs to be transferred with accuracy and low latency which in turn requires high bandwidth, especially with the increasing amount of mobile devices that generate most of today's traffic, mainly due to the evolving new mobile applications and use cases. In addition to that, the transport of data could be required in each stage of the Big Data architecture. This will raise a challenge in the transport and exchange of big data content both locally within the same stage of the Big Data architecture and non-locally between different stages of the same architecture. In both cases, the requirement for data content exchange could be the same in terms of bandwidth, latency, and accuracy.

BIG DATA DELIVERY

As noted earlier in this chapter, the size of the data being transmitted is growing and reaching a ginormous size, and a large proportion of this data is mainly related to the internet. The growth of the World Wide Web and new modes of Web services have generated an exponential growth in Web content and Internet traffic [5] & [6]. The Web consists of static contents (e.g. Static HTML pages, images, documents, software patches), streaming media (e.g. audio, real-time video) and varying content services (e.g. directory service, e-commerce service, file transfer service) [7].

Customers as well as companies and vendors, require fast web performance to satisfy their needs. Individual Web servers find it difficult to satisfy those needs of end-users. In order to store and serve huge quantities of Web contents, Web server farms - a cluster of Web servers functioning as a single unit - are presented [8] to address this high and increasing demand. However, these Web server farms are still finding it difficult to deal with flash crowds [sudden increase in simultaneous requests for a popular content (Major news Web sites experience this problem during major world events.)] which are frequently experienced in Web traffic [9].

Many solutions were introduced and much research has been invested to address the areas of the data processing, analysis, storage and in the field of securing data in order to meet the required hungry demand for it. Latency and web overloading is one of the fundamental research areas.

Due to the huge amount of available data failing to be transferred using the current communication systems, storage devices are still being shipped between offices and or data centres for the purpose of data exchange and delivery. Also, where possible another solution is used by compressing data before transmission.

A general strategy to deal with the delivery of the content is to replicate the content in one or more servers that are geographically close to the end-user, which represents a partial solution. However, it is of limited effectiveness for real-time applications. A better way for a real-time scenario is to use the distributed file systems, such as lambda and storm architectures.

Nevertheless, there are three main efficient ways for data delivery that could be used:

- **Web Caching**
- **Content Delivery Networks (CDNs)**
- **Peer – to – peer (P2P) Networks.**

The strategy is to direct the workload away from possibly overloaded origin Web servers. This is helpful because the clients' requests can be sent to the nearest and least-busy server. Moreover, value-added services, such as virus checking, Web page language translation, and automatic content adaptation for small handheld devices can be developed using this strategy.

The main advantages of the above mechanisms are:

- **reduce latency**
- **reduce network traffic, and**
- **save bandwidth**

Web caching addresses Web performance and scalability from the client side. Caching stores a copy of the data in a location close to the data consumer (e.g. in a Web browser) to allow faster data access than if the content had to be reclaimed from the origin server. The data will be stored in the cache once an original request for data has been successfully achieved. Any further requests for those files (a Web page complete with images, for example) will result in the information being reverted from the cache rather than the original server. In this way, the user will experience a very fast download since the request did not have to navigate the internet web. The latter will allow for all files to be served by a local source. In addition, the bandwidth that would normally be used to download the Web site is no longer required, and it will be free for other information delivery or transmission.

A Content Delivery Network (or CDN) is a system of globally positioned edge servers that delivers content on behalf of the original server. [10]

These edge servers maintain cached copies of the content (web pages, media files, etc.), and are distributed in different locations. In contrast to web caching, service provider's direct the user's client (browsers and native clients on PCs, smartphones, tablets, etc.) to use a nearby node for the server for accessing the content. The result is a reduced latency and faster load time due to data traveling a shorter distance. As the request will be delivered from the edge server's cache instead of the origin, this will reduce the load and improve scalability. CDN's can also deliver HD video, 4K content, as well as a multitude of other files. [11]

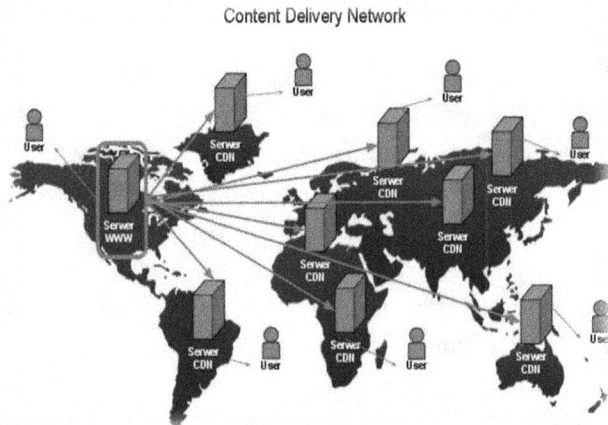

Figure (5). Diagrammatic Representation of a CDN, source: [12]

Questions regarding CDN operations include; how many and where to locate the cache nodes? How to place the content based on popularity? (The effectiveness of caching relies heavily on the existence of Zipf's law). CDN infrastructure constraints and content access patterns are other factors to consider for CDN operation.

Many companies are providing solutions for the content delivery, but the two main companies are AKAMAI as a market leader and GOOGLE.

The world's largest content distribution network, owned and operated by Akamai, spans more than 175,000 servers in more than 100 countries around the world [Akamai webpage]. However, since around 30- 40% of the total Internet traffic of an ISP (Internet Service Provider) points towards Google, the latter built its own infrastructure based on three distinct elements:

- **Core data centres**
- **Edge Points of Presence (PoPs)**
- **Edge caching and services nodes (Google Global Cache, or GGC)**

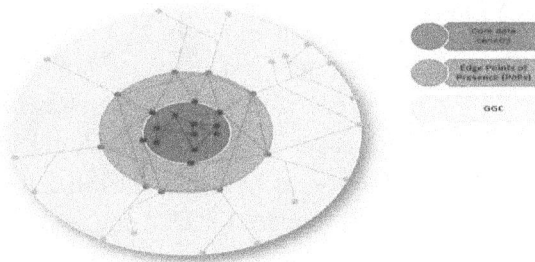

Figure (6). Google's network infrastructure, Source: www.google.com

The edge nodes (called Google Global Cache, or GGC) represent the tier of Google's infrastructure closest to the users. Static content that are very popular with the local host's user base, including YouTube and Google Play, are temporarily cached on edge nodes. Google's traffic management systems direct user requests to an edge node that will provide the best experience. [Google website]

Peer-to-peer networking (P2P) is a server-less, decentralised networking technology that allows several network devices to share resources and services and communicate directly with each other. [13] These resources and services include the exchange of information, processing cycles, cache storage, and disk storage.

Unlike the client/server model, P2P is typically an ad-hoc distributed network, where the addition and the removal of nodes have no significant impact on the network. The P2P network model allows each node to function as both a client and server, in which each party has the same capabilities and either party can initiate a communication session. Nodes can join and leave the network frequently, and that might not have permanent network (IP) addresses. An important factor in designing such a system is the location transparency problem.

One of the most important applications which relies on the P2P principles is the Voice over IP (VOIP), like Skype, which makes use of the unstructured, super – peer method. Another example of the P2P pattern is the use of Distributed Hash Tables (DHT) that has been adopted by Amazon's Dynamo and Facebook's Cassandra.

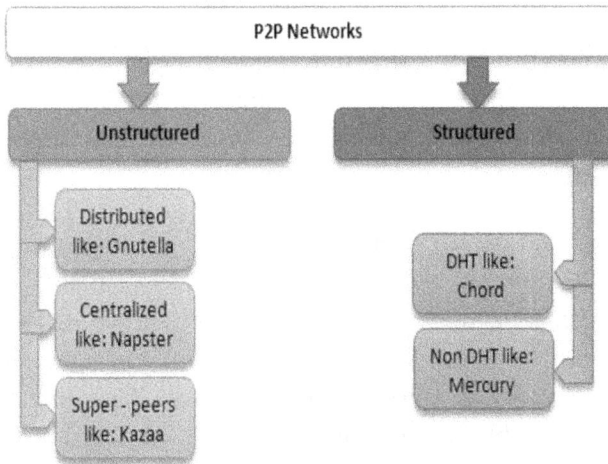

Figure (7). Classification of P2P Networks

5G AND BIG DATA

Data delivery is an area of challenges for the mobile network industry due to the increasing demand for mobile broadband. This increased demand is encouraged by the rapid advances and growth in the number of devices attached to these networks.

The software applications' industry is highly demanding and profitable. The total expected revenue was projected almost to double between the years 2015 and 2017 to $76.5 billion (http://www.businessofapps.com/app-revenue-statistics/).
Applications are designed for different uses, starting from the real needs and usual developments to entertainment and social applications, passing through the medical and environmental applications that are in direct or indirect touch with people's lives. This competitive and successful money-making market is

continuously increasing the number of new software applications presented. There are vast ranges of technology developments on the horizon, and most of these developments are based on the higher demand for data and low latency connected to the term of IoT in one way or another.

It is predicted that the IoT will be the result of the interworking of M2M, Virtual reality, intelligent or smart society, in addition to the development and expandable use of a sensor network and the new wwww (worldwide wireless web). IoT will generate massive volumes of data (Big Data) and will require huge scalability and reliable connectivity.

The mobile/wireless form of communication is becoming the foundation for many industries, including M2M, sensor networks, business-process optimisation, and consumer electronics. It is highly dependent on mobile devices, remote applications, radio capabilities and the integration of different and various networks.

All the above leads to one main question; is the mobile/wireless communication industry and available technology ready to handle Big Data?

In 2012, ITU-R embarked on a programme to develop "IMT for 2020 and beyond", which provides the framework for "5G" research and development activities worldwide. In September 2015, ITU-R finalized its "Vision" of the "5G" mobile broadband connected society. [14]

So what is the 5G vision and what are the requirements and use cases for 5G?

5G will be built to integrate networking, computing and storage resources into one programmable and unified infrastructure.

This new model of the unified system will allow for an optimised usage of all distributed resources, and the convergence of fixed, mobile and broadcast services. In addition, 5G will enable prioritisation of media content and will support multi-tenancy models, enabling operators and other players to collaborate in new ways. 5G is the door opener for new opportunities and use cases, many of which are as yet unknown. In 5G, everything is expected to be connected, with unlimited access to information and the ability to share data between the connected things without boundaries which will represent a challenge to the new generation of mobile networks. Nevertheless, 4G is just in its beginnings, and its capabilities will continue to develop to support many new use cases and applications. Hence many of the features that could be seen in 5G may, in fact, be implemented as LTE-Advanced extensions and enhancement before the full release of 5G.

The parameters that 5G technology will be developed upon include: [15]

- **Traffic capacity;**
- **Data throughput;**
- **Data integrity;**
- **Latency;**
- **Energy consumption;**
- **Technology convergence; and**
- **Smart communication.**

5G will be required to satisfy most of these demands and use cases, which in most cases are unpredicted. A 5G task will most likely be the integration of most of the existing technologies and the broad, diverse types of devices connected to these technologies. In the predecessor generation, most of the communication was to take place between people through the devices attached to the network.

5G communication is expected to take place between devices that will form the Internet of Things (IoT). This will in turn enable the Machine Type Communication (MTC) which will require higher demands for bandwidth, network availability, and low latency.

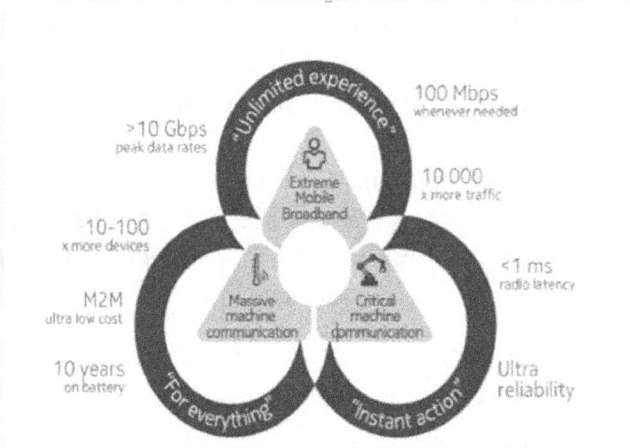

Figure (8). Heterogeneous use cases – diverse requirements [16]

Some terms could be seen impossible or very difficult to achieve. However, for example, "Zero latency" does not necessarily mean there will be no delay at all, it simply represents the fact that latencies should be low enough in 5G so that it will be able to fulfil the needs of the most challenging use cases, such as Virtual reality.

Potential expected requirements for the 5G can be summarised as below: [17]

- **1-10Gbps connections to end points in the field (i.e. not theoretical maximum)**
- **1-millisecond end-to-end round trip delay (latency)**
- **1000x bandwidth per unit area**
- **10-100x number of connected devices**
- **(Perception of) 99.999% availability**
- **(Perception of) 100% coverage**
- **90% reduction in network energy usage**
- **Up to ten-year battery life for low power, machine-type devices**

It can be noted that data, latency, and bandwidth will be the key drivers for 5G since users will always require more data due to the exponential increase of the everyday new, different applications. Therefore it is necessary to find solutions for many of the use cases that require such needs.

From the viewpoint of IMT 2020, the key constraints are set by the following two submission deadlines:

- **Initial technology submission by ITU-R WP5D Meeting #32, June 2019**
- **Detailed specification submission by ITU-R WP5D Meeting #36, October 2020**

CONCLUSION

A key factor for any problem solving is to divide the problem into small parts and try to find a solution for each part, as this will reduce complexity. In a similar way to this, the best way to deal with big data is to divide it into stages or levels and try to find solutions for each stage or level separately. These solutions could be combined and/or duplicated to build an efficient system that can extract the value hidden inside the data. This will also allow managing the solutions which will potentially lead to better decision making in terms of cost and time reduction and development of new services.

One of the most important parts to all of this is the management and delivery of the Big Data which represents a fundamental element in developing the upcoming 5G network, which is a foundation Infrastructure for IoT.

REFERENCES

[1] Steven B Andrews and David Knoke. Networks in and around organizations. Jai Press, 1999.
[2] Big Data Insights, http://www.sas.com/en_us/insights/big-data.html
[3] Sebnem Rusitschka, Alejandro Ramirez (SIEMENS): Big data roadmap and cross-disciplinarY community for addressing socieTal Externalities, 01 March 2014. www.byte-project.eu
[4] Gartner, IT Glossary, Big Data: http: http://www.gartner.com/it-glossary/big-data
[5] Rabinovich, M., Spatscheck, O.: Web Caching and Replication. Addison Wesley (2002)
[6] Molina, B. et al., A closer look at a Content Delivery Network implementation, Melecon 2004, Dubrovnik, Croatia, 12-15 May 2004
[7] Rajkumar Buyya, Mukaddim Pathan, and Athena Vakali (Eds.), Content Delivery Networks, Lecture Notes in Electrical Engineering (LNEE), Vol. 9, Springer-Verlag, Germany, 2008. ISBN: 978-3-540-77886-8
[8] R. Burns, R. Rees, and D. Long. Efficient data distribution in a web server farm. IEEE Internet Computing, 5(5):56 – 65, September - October 2001.
[9] C. Pan, M. Atajanov, T. Shimokawa, and N. Yoshida. Design of

adaptive network against flash crowds. In Proc. Information Technology Letters, pages 323–326, September 2004.

[10]https://www.akamai.com/us/en/cdn/

[11]Beatriz Grafulla-González, Senior Researcher, Ericsson: Content Delivery Network Optimization, 2016.

[12]http://www.wpmayor.com

[13]https://technet.microsoft.com/en-us/library/bb457079.aspx

[14]http://www.itu.int/en/ITU-R/study-groups/rsg5/rwp5d/imt-2020/Pages/default.aspx

[15]5G: What is it, Ericsson OCTOBER 2014. www.ericsson.com

[16]5G masterplan – five keys to create the new communications era, Nokia

[17]Dan Warren, and Calum Dewar, Understanding 5G: Perspectives on future technological advancements in mobile, GSMA Intelligence, 2014.

[18]Jose Costa-Requena, Raimo Kantola, Jesús Llorente, Vicent Ferrer, Jukka Manner: Aalto University, Aaron Yi Ding, Yanhe Liu, Sasu Tarkoma: University of Helsinki, "Software Defined 5G Mobile Backhaul", EIT ICT, CELTIC Plus SIGMONA project.

[19]Content Delivery Network Optimization, Ericsson. www.ericsson.com

[20]Choon Hoong Ding, Sarana Nutanong, and Rajkumar Buyya: P2P Networks for Content Sharing, the University of Melbourne, Australia

[21]Markus Hofmann and Leland R. Beaumont: Content Networking: Architecture, Protocols, and Practice, Morgan Kaufmann Publishers Inc., San Francisco, CA, USA, 2005.

CHAPTER FOUR

Boosted Prediction Analysis for Big Data

Hayder K. Fatlawi

Information Technology Research and Development Center

The University of Kufa, Najaf, Iraq

ABSTRACT

The optimal management of healthcare's resources contributes to improving the quality of medical services and thereby enhancing the level of the health of society. This management requires providing prospective information about the patient admittedto a hospital. Prediction techniques represent a useful tool for knowledge discovery in a massive and complex healthcare dataset. In this chapter, a prediction model has been designed and implemented which analyzes medical records of patients and provides information for decision making in health institutes. The proposed model consists of three primary stages; the first stage is preprocessing data that focuses on preparing the information for the mining process. It includes data binarization, attribute construction and data rows aggregation. These steps make healthcare data more suitable for the prediction technique. In the second stage, a model is developed, Gradient Boosting

Machine (GBM) algorithm, which reduces the prediction error by building a sequence of binary regression trees. GBM overcomes some weakness points in forecasting techniques that depend on a single tree. Improvement has been performed on GBM by replacing typical splitting criteria by correlation measures in regression tree building. Another improvement is using the fast formula for choosing the best split point. Model evaluation in the third stage has been performed depending on popular measures of error prediction and cross-validation. The real and massive dataset is evaluated by RAE, MAE, MSE, RMSE log measures with a three folds cross-validation on the tree to ensure precision and reliability of evaluation. Three methods for testing the final model are used; (i) single last tree (ii) random choosing of the trees (iii) combination of all trees.The experimental result shows that the proposed model which depends on correlation has better results when compared with a typical GBM. The model with 160 regression trees has shown a testing error of 0.468 in RMSE log, while the measurement error of the original GBM was 0.49. Training time is reduced by 85%, going from (1504) minutes to (254) minutes.

Keywords: Prediction Modeling, Data Mining, Big Data, Healthcare.

INTRODUCTION

Healthcare concerns applying all necessary medical procedures to restore the health of people or prevent aggravation of health problems [1]. The cost of those proceedings are growing rapidly to satisfy the suitable quality of health care systems and this raises financial problems. Furthermore, the increasing advancements in this field make the implementation of techniques more sophisticated. There are many tools used to deal with the challenges of healthcare issues such as statistical methods, artificial interlining, and data mining techniques. Data mining is gaining useful knowledge from machine learning techniques and statistical methods [2]. It has three main phases. Firstly, the data is prepared for the mining process, then machine learning techniques for knowledge extraction are applied. Finally, the results will be developed in an understandable form to aid us in decision making. Data mining is mainly used for two purposes, discriminatinghidden patterns in data and predicting unknown values of interested features from known values of other elements [3]. This work focuses on prediction techniques of data mining. Methods of prediction are used widely to solve health care problems thereby reducing costs [4]. These methods allow the forecasting of some future information such as predicting hospitalization numbers for the coming years based on historical records of patients. Furthermore, they provide models to specify which admission is unnecessary as well as making a clear point regarding the future requirements in health care systems. In resemblance to other medical data sets, the data sets of health care have many properties which make the prediction process more complicated [5]. In this work, we introduce a predictionmodel for predicting the future hospitalization of patients using ensemble machine learning and a regression model.

The predictive analysis could be defined as the task of data analysis to predict unknown values of the prediction target feature. It includes a classification task for class label prediction and a numerical prediction where the task is to predict continuous values or ordered values. The type of target attribute specifies if the problem is classified as a binary value or a numerical prediction with constant values. Many statistical methodologies are used for numerical prediction such as regression analysis, which is the most commonly used method [6]. Regression techniques are used with numerical prediction. They vary according to the complexity of data of the prediction process. The relation between a target attribute and other interested attributes specifies how difficult the prediction will be. With a linear relationship, we can use a technique like linear regression. On the other hand, more advanced techniques are needed for more complex relations [7]. A decision tree (DT) is an easy and useful tool for binary class problems. It utilizes a recursive binary splitting to build a prediction model. It is developed to handle regression problems when multiple values for the target variable are founded. The decisionthat is made by a single regression binary tree may not be precise especially with using high complex data [8, 9]. Boosting is a machine learning method for building a strong predictor from a set of weak predictors to overcome the problem of bias in the prediction model. A Gradient Boosting Machine (GBM) is one of the best implementations of boosting regression trees, and it depends on the concept of reducing the training error by building a sequence of trees [10, 11, 7]. Figure (1) summarizes the types of prediction techniques and boosting methods.

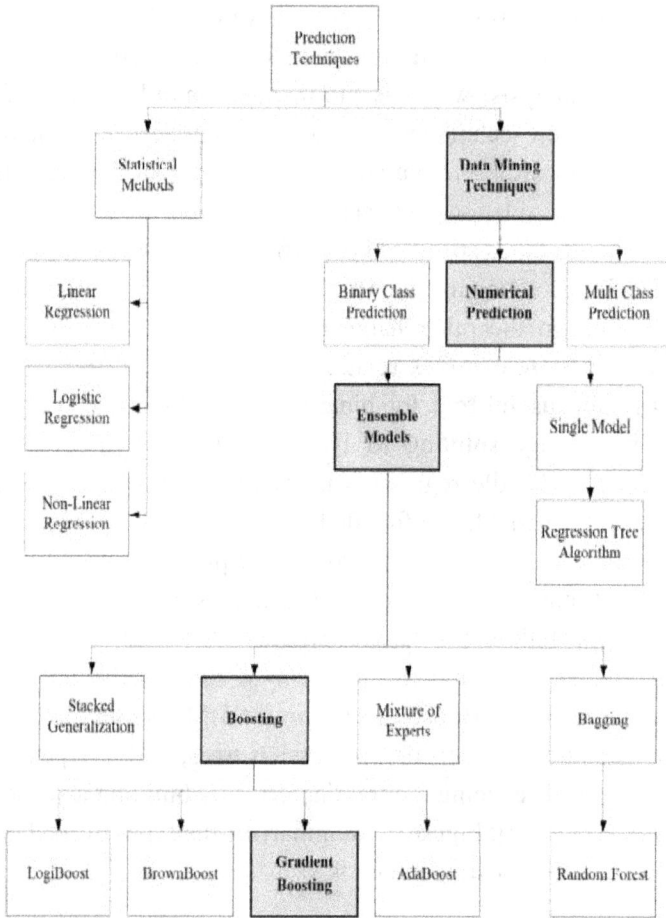

Figure (1) Distribution of Prediction Techniques.

Data mining techniques have been applied in different kinds of research, in hopes of solving the problems concerning healthcare. Related works in this field are various ranging from recommended systems and prediction models for unnecessary hospitalizations. These operations will be reviewed in this section, and the points of similarity and dissimilarity will be explained. Dimitris et. al., 2008 [12] utilized modern data mining methods for healthcare resource problems by collecting many sources of claims data and grouping them using classification and clustering with some performance measures such as hit ratio, penalty error, and R^2. Dimitris's work focused on data preprocessing and used typical data mining techniques for the domain of health care. On the other hand, our thesis focuses on improving a prediction model for the same area in addition to data preprocessing. While, Xiang et. al., 2011 [13] used machine learning algorithms to reduce unnecessary hospitalization by using records from the previous year to predict how long the patient will stay in the hospital for the coming year. Researchers utilized the support vector machine (SVM), random forest, regression tree and boosting ensemble with the first release of the Heritage Provider Network (HPN) Dataset. Xiang dealt with a healthcare dataset that uses multi-class and regression techniques. Our thesis focuses on developing a GBM prediction method with the same dataset but with a third release, instead of using a typical GBM. Rashedur and Fazle, 2011 [14] used and compared different decision tree classification techniques to classify admitted patients according to their critical condition. They developed an application to diagnose and measure the criticality of the newly arrived patient through the use of a mining hospital surveillance unit using a C4.5 Decision Tree Classifier. Afterward, they evaluated results using a False positive rate (FP), Recall and Precision. Our experiment is similar to this in the domain of healthcare while differs by using a different dataset, evaluation measure and regression (rather than classification). Hadi and Nima, 2012 [15] proposed the correlation of a splitting criterion to a multi-branch decision tree.

They replaced the traditional approaches for building the decision tree (Gain ration and Gini index) so that the attribute that has the largest absolute correlation with the target is chosen as the best splitter in each node. Furthermore, they used the ability of SVM to maximize the margin between classes and hence determine the best threshold on the selected feature. The proposed work is close to this work in using the correlation as a splitting criterion, but it is utilized for building binary regression trees of GBM instead of the traditional standards (sum of squared error and standard deviation).

Moreover, we use this correlation for the predictive task rather than the classification task. Also, in our approach, a simple method for finding the best threshold rather than using SVM is founded. Jufen et. al., 2013 [16], constructed a Chi-Squared Automatic Interaction Detection (CHAID) classification tree with a 10-fold cross-validation to predict the probability of death or hospitalization for heart failure patients. He later compared the result with logistic regression (LR) models using a Receiver Operating Characteristic (ROC) curve analysis based on a TEN-HMS Dataset. It was deducted that the CHAID tree performed better than the LR-model in predicting the composite outcome. Our thesis is close to Jufen's work due to the utilization of the same task of data mining which is used for prediction in healthcare situations. While our research varies from that of Julien's regarding using a different dataset and technique for prediction. Nannan, 2014 [17] compared the performance of three prediction methods (i.e., linear regression, random forest, and gradient boosting) on hospitalization dataset to discover which technique is the best. By experiments, Nannan came to the conclusion that the random forest method provides the best prediction of patient admission. His work used typical prediction techniques while our thesis aims to develop a new prediction model for the same dataset.

Predictive Data Mining: Concepts and Tools

Data mining is defined as "extracting or "mining" knowledge from large amounts of evidence [3]. Another definition for data mining is "the automatic discovery of previously unknown patterns or relationships in a vast and complex dataset" [18]. In data analysis, working with data commonly includes descriptive and predictive tasks. A descriptive analysis could be used to visualize a relationship of information in an understandable form for decision makers. Predictive analysis of data is a little different from the illustrative use of data. In a predictive task, it is okay to describe a relationship among data elements and a prediction model will be used to approximate reproducing the relationship with new information [8].

A prediction task is a two-step process, consisting of a learning model step (i.e. a prediction model is constructed) and a prediction step (i.e. the model is used to predict a target attribute for given data) [2, 6]. The input data for a prediction task is a collection of records. Each record is represented by a tuple (x, y), where x is the attribute set and y is the target of prediction. The type of target attribute y detects the essential technique. If values of y are binary, binary class classification methods can be used. While multi-class classification is utilized if y has more than two categories. If values of y are continuous, regression techniques should be employed for the predictive task [2].

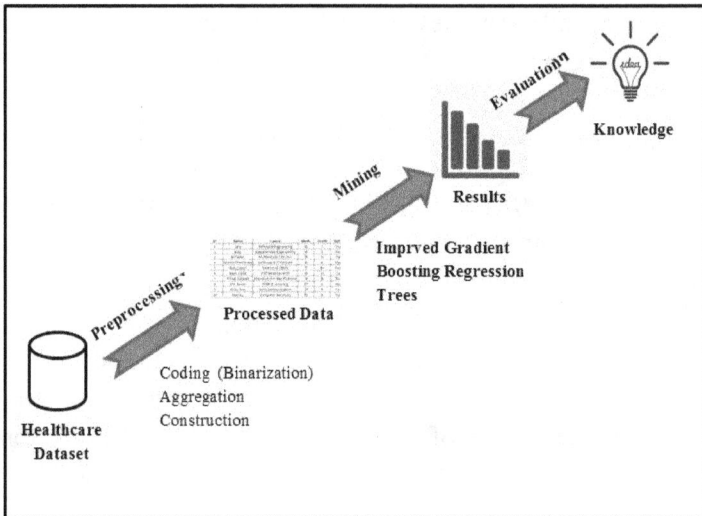

Figure (2) KDD process stages according to the scope of Work

Data Preprocessing

For the most part, real-world databases tend to contain low-quality data which cannot be used directly in the mining process without employing the preprocessing step [3]. Data preprocessing techniques are classified into two groups. The first group is concerned with cleaning the data from noisy, missing, duplicate, and inconsistent data samples. The second panel focuses on reconstructing the data by binarization, constructing the attributes and aggregating data rows [3]. This section explains the concepts of the second group of preprocessing in detail. These methods improve the efficiency of the mining process and reduce the complexity of the required resources (i.e. storage and time).

Data Binarization

Data binarization is one of the transformation methods which aims to produce a new set of attributes from the original one. It includes transforming a single categorical attribute, *x,* that has *m* categories, into *m* of new attributes *(x1-xm)*. The values of new attributes are binary, the value of *x1* will be one if x = x1 and 0 if x≠x1 [9].

Data Rows Aggregation

Aggregation combines multiple data objects into a single object. The major advantage of the combining technique is the resulting reduced data sets. Since data reduction requires less memory and less processing time, this allows the use of more expensive data mining techniques [2]. Numerical attributes, such as a counter of surgeries, are aggregated. This can be done for all hospital visits by applying some statistical operations such as sum, average, maximum, minimum, and standard deviation.

Attributes Construction

The original set of attributes may not be useful for mining instantly, so many methods are applied to produce a new round of attributes. Quality construction is one of these methods, and it includes constructing new attributes from original attributes. It aims to improve the accuracy and understanding in high-dimensional data. For example, the mining process may concern the area information, and we have information about the height and width attributes only. In this case, construction of area could be applied. Quality construction can provide knowledge discovery by discovering hidden information about the relationships among data attributes [3].

Binary Class Prediction (Decision Tree)

A decision tree is a simple and powerful form of data analysis which allows us to predict, explain, describe, or classify a target. [8]. A decision tree is represented by a flowchart-like tree structure, where every nonterminal node denotes a condition on an attribute. The tree is used to split data records which have different characteristics. Each branch represents the result of that condition, and each leaf node (i.e. terminal node) holds a class label. The first node in a tree is the root node. In a binary decision tree, each internal node branches to exactly only two other nodes [2, 3]. For a new data record X, which has an unknown target classy, the values of attributes of X could be tested against the decision tree. Tracing starts from the root node to leaf node, which assigns the value for a class of X. The building of decision tree classifiers does not require any domain knowledge, and therefore is suitable for exploratory knowledge discovery. Representation of acquired knowledge by using a binary tree form is self-evident and easy to understand by humanity [6]. For healthcare, binary classification may be utilized for many tasks such as the possibility to visit a hospital. A decision tree can be used to analyze historical records of the patient and predict whether that patient will visit the hospital in future. So the target has two values only (admit and not admit). The learning step will try to choose the most related attributes for splitting records during the building of a decision tree. Prediction on a decision tree can be performed by starting at the root of the tree and moving down through internal nodes until a leaf node is reached, which will provide the value of prediction. Figure (3) gives a simple example of binary class prediction by using decision trees.

Example training healthcare data

No	Age	Gender	Admit
1	51	Female	Y
2	12	Male	N
3	10	Male	N
4	13	Female	N
5	25	Female	Y
6	28	Male	Y
7	31	Female	Y
8	12	Male	N
9	8	Male	Y
10	63	Male	Y

Where Y: admit, N: not admit

New record with unknown target

| 11 | 19 | Male | ? |

Path of prediction:

Node 1 → Node 2 → Node 4

Admit = N

Root

No. record = 10 Node1 Age<=25

Yes No

Internal Node No. record = 6 Node2 Gender=Male

Node3 Terminal Node No. record = 4 Target: Y

Yes No

Node4 Node5

Terminal Node No. record = 4 Target: N

Terminal Node No. record = 2 Target: Y

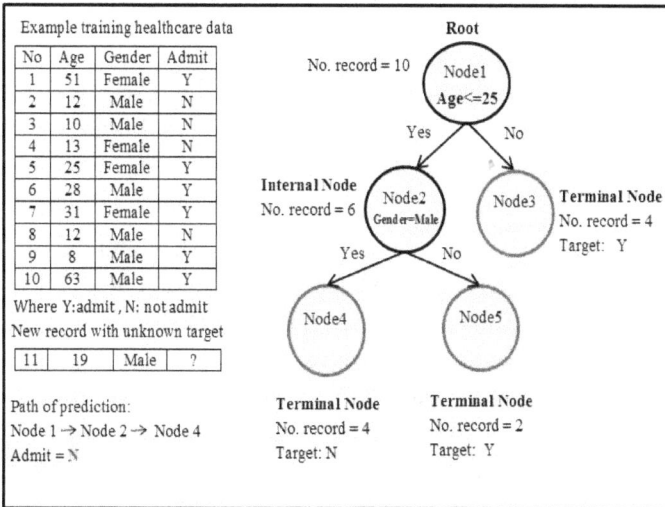

Figure (3) Binary class predictive task for healthcare data

Multi-Class Prediction

If the target of the prediction function has k possible values, where k > 2, techniques of classification should be extended to allow for a multiclass prediction task. One-versus-all (OVA) is a simple approach which deals with k classes by training k binary classifiers, one for each class [6]. As an extension of the health care predictive task which was mentioned in section (2.3.1), the target may have three possible values for the duration of hospitalization (Long, short, and not admit). Hence, three classifiers could be used to solve this task. First, a classifier is trained for value Long as a class, and the other values (short, and not admit) would be treated together. The other two classifiers could be formed in the same manner.

Numerical Prediction

The term forecast includes both numeric prediction and class label prediction. While the binary and multi-classification techniques predict categorical (discrete, unordered) class names, regression techniques model continuous-valued functions. That is, regression is used for predicting unknown numerical data values instead of distinct class labels. Regression analysis could be defined as a statistical methodology which is mostly used for numeric prediction [19, 6]. In this thesis, continuous prediction values that are focused on in healthcare data analysis are studied and the question, "How many days will be spent by a patient in the hospital next year?" is answered.

A regression tree is an extension for a typical decision tree and is used for performing the numeric prediction tasks. Each leaf (terminal node) stores the average value of the data records of target Y. This stored information is what was accumulating inside the leaf during the training process. For predicting any new data records, a regression tree is followed down to a leaf using the file attribute values. When data reaches a terminal node, the tree assigns the previously stored value of target Y to that record. Mostly, a decision tree chooses the splitting attribute to maximize the information gain. For numeric predictions, it is appropriate to minimize the variation in the target values in each node [20, 21]. Less change means more homogeneity in which the values become close to each other and that will reduce the error of prediction. A regression tree is growing as follows [7]: data contains n records, each one consists of m attributes and a target. Each record has a tuple (x_i, y_i) where $i=1, 2,...,N$, and x_i = $(x_{i1}, x_{i2}, . . . , x_{im})$. Afterward, the regression tree should decide which attribute will be used for splitting data at the root of the tree and what is the splitting point? (i.e. age < 20). A binary regression tree (BRT) starts with producing two child nodes from a root that will split the data into two regions R1, and R2. For this level of tree, the target can be modeled as a function f(x), as follows:

$$f(x) = \frac{\sum_{i=1}^{N_{R1}} y_i}{N_{R1}} I(x \in R1) + \frac{\sum_{k=1}^{N_{R2}} y_k}{N_{R2}} I(x \in R2) \quad (1)$$

Where N_{R1}: number of records belonging to R1, N_{R2}: number of documents belonging to R2. $I(x \in R1)$ is a binary function that detects if record x belongs to region1 (first child node) or not, and the same with $I(x \in R2)$, which detects if x belongs to region2 (second child node) or not. This procedure is repeated recursively for each child node to produce new child nodes in which R1 will be split into R3, R4 and R2 will break into R5 and R6. Each split operation is required for choosing an attribute as a splitter and choosing the best split point to get more homogenous target values.

The quality of an attribute and the quality of a particular split point for dividing the data are measured by the sum of squared error (SSE). The goal is to minimize SSE between target values of a data record in a particular node, and the average of the target of that node in BRT is measured [22]. The accuracy of a split operation that divides the data into two regions R1, R2 can be estimated by the following equation:

$$SSE = \sum_{yi \in R1}(yi - \overline{y1})^2 + \sum_{yi \in R2}(yi - \overline{y2})^2 \ldots\ldots (2)$$

Where $(y1)^-$ and $(y2)^-$ are the averages of the target values in regions R1 and R2, respectively. BRT chooses the attribute and the split point which minimizes SSE to a small point. Figure (3) illustrates the evaluation of the attribute's efficiency for a particular split point, for one level of BRT.

Example of training
healthcare data

No	Age	Gender	Admit /days
1	51	Female	6
2	12	Male	0
3	10	Male	0
4	13	Female	0
5	25	Female	1
6	28	Male	0
7	31	Female	2
8	12	Male	0
9	8	Male	1
10	63	Male	9

Node1
Age<=25

No. of records = 10
SSE = 86.9

Yes   No

Node2   Node3

No. of records = 6   No. of records = 4
SSE = 1.33333   SSE = 48.75

Goodness of split point (25) in attribute (age) = SSE before splitting – SSE after splitting

$$= 86.9 - (1.33333 + 48.75)$$
$$= 86.9 - 50.083$$
$$= 36.817$$

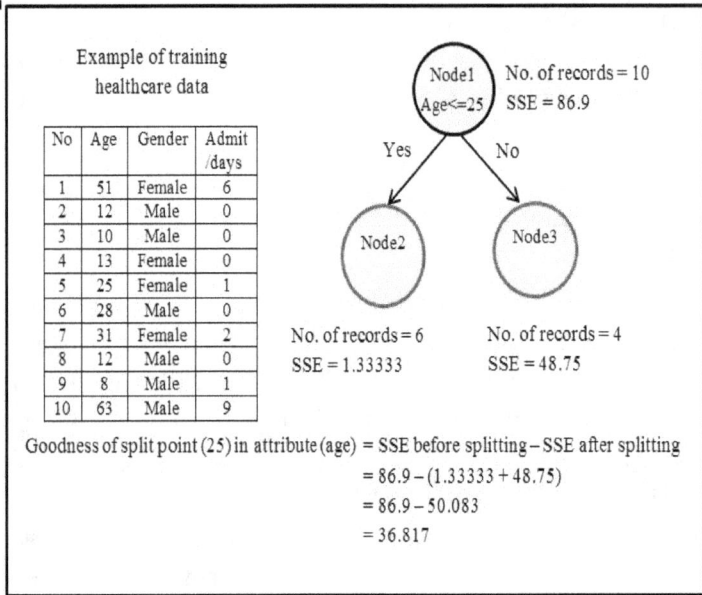

Figure (4) Evaluation of attribute's goodness during BRT building

Standard deviation STD is another splitting criterion that is used to find the quality of an attribute (and split point). The value of STD represents an indicator for homogeneity of data before and after the splitting operation. If the overall sum of STD in child nodes is less than the STD of the parent node (more close to the zero), the splitting is preferred. Choosing splitting attributes (and split point) depends on which one has more reduction after cutting [23, 22]. A split operation divides the data into two regions R1 and R2, and its accuracy can be measured by the following equation:

$$STD_{Reduction} = STD(R) - \left(\frac{n_{R1}}{n} \times STD_{R1} - \frac{n_{R2}}{n} \times STD_{R2}\right)...$$
(3)

Correlation Criteria

Correlation can be defined as a measurement of association among data. If values of predictions of target Y increase when values of attribute x increase, the relationship would be called a positive correlation. The term negative correlation is used when values of Y decrease after increasing values of x. It is usually utilized for attribute selection in pre-processing. There are many types of relationships and selecting one depends on the nature of data. One of the most popular kinds of correlations is Pearson's correlation. Its equation is as follows [24, 25, and 26]:

$$wj = \frac{\sum xjY - \frac{\sum xj \sum Y}{N}}{\sqrt{\left(\sum xj^2 - \frac{(\sum xj)^2}{N}\right)\left(\sum Y^2 - \frac{(\sum Y)^2}{N}\right)}}... \quad (4)$$

Where wj: a relationship between attribute xj and target Y, $\sum xj$: summation of values of attribute xj, $\sum Y$: summation of values of the target.

In this proposed work, correlation replaces typical criteria that are used to choose the best quality for splitting data. After the best attribute is determined, the best split point from all possible points should be selected. The typical method is to try all possible split points and evaluate the quality of each point according to the formula that was presented in the previous section. This operation is a costly computational task because the effort increases as the values of attributes increase. For example, with a continued quality which has k values, there is the k-1 possible split point. Luís Fernando proposed a simple measurement for choosing the best split point as the following formula suggests [27]:

$$Err(sp) = \frac{Sum_L{}^2}{N_{tL}} + \frac{Sum_R{}^2}{N_{tR}}\ldots \quad (5)$$

Where $Err(sp)$: Error after splitting data based on split point sp, $SumL$: summation of target values on a left child node, $SumR$: summation of target values on a right child node, NtL: number of data rows in the left child, NtR: number of target values on the right child node.

Ensemble Model

Models based on a single tree have some weakness points, such as; (i) model instability, which means any slight change in data may alter the structure of the tree, (ii) larger prediction error in single **BRT**. As explained, splitting the data should produce rectangular regions that contain similar target values. If the relationship between the selected attributes and the target cannot be sufficiently defined by rectangular areas, then **BRT** will have a larger prediction error. To solve these issues, ensemble methods have been developed which combine many trees [22].

Ensembles have more power predictive performance than a single tree, so they became the preferred techniques for both classification tasks and numeric predictions [20, 22].

The methodology of an ensemble model involves combining a set of single models, each working to solve the same original task, in hopes of obtaining an enhanced integrated global model [30, 31]. Two points should be taken into account when using ensembles; (i) the size of the group (ii) the mechanism of combination among the results of trees [28].

Many techniques are developed for a group model. Such methods include bagging, boosting, and stacking. *Bagging* combines the decisions of multiple trees by using a voting concept for binary (and multi) class predictive tasks. When employed in mathematical predictive functions, bagging calculates the average. A modern example of the bagging techniques is the random forest.

Boosting adopts the same approach but with the significant difference that bagging gives data records equal weight in all the models while in boosting, the data files that have not been predicted correctly by the previous model will have more weight [20, 29, 30]. That means each type of data record involved in boosting is influenced by the behavior of previous models.

Boosting

For numerical prediction tasks, expanding is considered as "a statistical optimization technique that aims to minimize the loss in predictive performance of the model" [32]. Boosting motivation combines the results of many "weak" prediction models to produce a robust model [7]. With this technique, **BRT** models are made to fit the training data iteratively, by using suitable methods to gradually increase the focus on data records that have been incorrectly predicted. Boosting algorithms are different regarding measuring the lack of fit and choosing parameters for building the model [32].

The loss function, such as a squared error, is reduced by repetitively adding a new decision tree that makes its way down the gradient of the error. The first BRT maximally reduced the loss function and with the carrying out of iterations steps; it focuses on the residuals in the target's values [32].

Gradient Boosting Machine

Investigating the statistical framework of boosting led Friedman to propose a gradient boosting machine (GBM) algorithm. It is a simple, elegant, and highly adaptable technique for both class and numerical predictive tasks [22].

The gradient boosting machine is a powerful and brilliant prediction method which utilizes the promoting concept for reducing the error of prediction. GBM algorithm builds an additive model for minimizing the **residual value** which represents the value of subtracting each target value and the mean of the target [22]. *The target value for each data record is re-estimated at each iteration. The aim of that is to give the new tree its contribution* [32].

The linear combination of many binary regression trees represents the final model of GBM. The performance of the building process is best if it proceeds slowly. For this reason, the contribution of each **BRT** is reduced by a learning rate which has a value of less than one. The results of GBM are usually much more stable and precise than those of a single **BRT** model [32]. The gradient boosting machine that utilized **BRT** for learning requires four parameters [32, 7 and 22]. Those parameters have a major effect on the performance and behavior of GBM.

- The first parameter is *the maximum number of trees* in the model. This means how many trees would contribute in the boosting process. The importance of this parameter could be clarified with significant data (e.g. healthcare data) because it controls the balance of model complexity and prevents an extensive training process.

- The second parameter is the learning rate called **shrinkage**. As mentioned previously in this section, boosting is improved with a slow process, and reduction is used for this purpose. The contribution of each **BRT** in the final model is shrunk for the purpose of reducing the probability of overfitting [7]. Overfitting may happen when we fit the data in the training process too well, which increases the risk on future predictions [2, 3]. A small value of shrinkage is preferred. Usually, the best result appears when the value is less than 0.1 (decrease < 0.1) [10, 7]. Reducing the shrinkage value requires increasing the value of the maximum number of trees (the first parameter) [32].

- The third parameter is the Maximum number of terminal nodes in **BRT**. All binary regression trees in GBM have approximately the same size. The reason for this is to avoid overfitting. Moreover, it improves the efficiency of the algorithm from a computational viewpoint. A large tree needs more resources (time and memory) than a smaller tree. Furthermore, the ability of understanding and interpretation is one of the decision tree advantages and increasing the nodes could reduce this benefit [7].

- A fourth parameter is a minimum number of data records in a terminal node. This parameter represents a stopping condition for splitting in the **BRT** building process. When reaching to a node that has data records less than a particular value, the splitting will be stopped, and we need to consider that node as a terminal node. A small number of this parameter produces an extensive **BRT** and vice versa [32]. With consideration to the vast amount of healthcare data and its complexity, GBM introduces a reliable and accurate prediction tool. Figure (4) illustrates the working strategy of GBM with an example of healthcare data which is used in Figure (3).

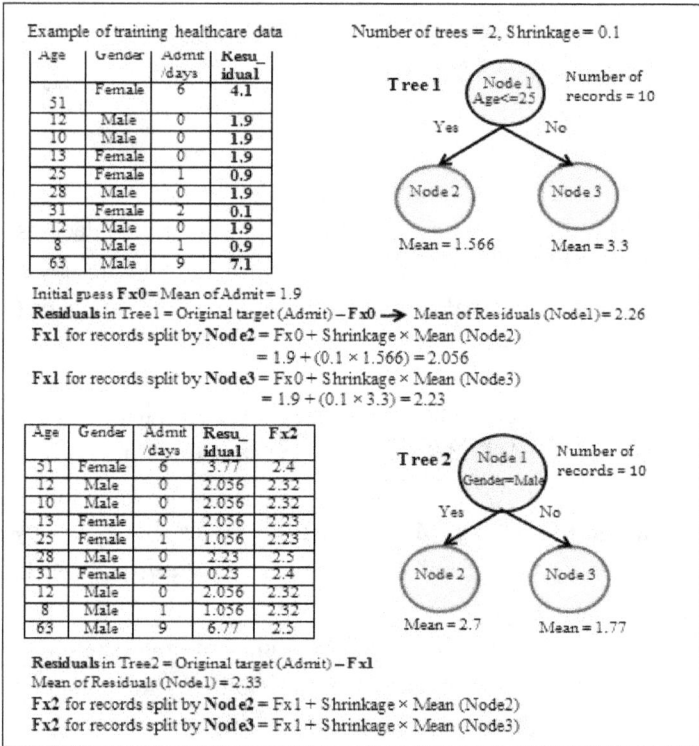

Figure (5) Gradient boosting machine based on two BRT and model of

In Figure (5), we summarized the idea of GBM with an example of healthcare data. The parameters of GBM are set as follows; a maximum number of the tree is 2, shrinkage is 0.1, a maximum number of terminal nodes is 2, and a minimum number of the samples is 1. An example of the data contains 10 data rows with two attributes and a target. The GBM method starts with calculating an initial guess (i.e. mean of target values for data rows). The average of the admitting days was 1.9, and we called it Fx0. The next step was to find the residual between Fx0 and every record. Those residual values are considered as a target for building Tree1. With the assumption that the best attributes in the root of Tree1 were Age, and the best split point was 25, the binary division produced two nodes. Node 2 within the data rows had aged less or equal to 25, Node3 had aged more than 25. After reaching the stopping conditions with two terminal nodes (i.e. Node2 and Node3), the building of Tree1 was stopped. The contribution of Tree1 was represented by Fx1, which was used to calculate two nodes. Fx1 for Node2 was produced by the summation of Fx0 and the result of various shrinkages by the mean of data rows in Node2. In the second iteration, the residual between the original target values and Fx1 was calculated and used as a target for Tree2. The best split point of the root of Tree2 was Gender Equal Male and the splitting also produced two terminal nodes. Fx2 was calculated based on Fx1 and the mean of the terminal nodes in Tree2. After the building of Tree2 was finished, the stopping condition of a maximum number of trees was reached, and the model of GBM was completed. This example explains the mechanism of building a boosted model based on regression trees and the algorithm of those steps will be described in detail in Chapter Three.

Other Boosting Techniques

There are other boosting techniques such as Ada Boost, Brown Boost, and Logi Boost [33, 29]. Ada Boost is a widely used algorithm that works on promoting a concept. It is designed for working with binary and multiple-class prediction tasks. Ada Boost starts with giving equal weight to all data records then commands decision tree algorithms to build a classifier for current data and then reweighs each data file according to the tree's result. The weight of correctly classified data records is decreased, and the weight of misclassified ones is increased. This method creates a set of secure data records having light weight, as well as a set of reliable information records having high pressure [7, 23].

Evaluation Measures

All binary regression trees that are built by GBM are evaluated based on test datasets that have never been seen before. MAE, MSE, RAE, RMSE Log and measures are used to assess the prediction error of every tree and then find the total error for a combination of all the tree models. The equations and steps that are mentioned above could be described as follows [13, 22]:

- Mean Absolute Error (MAE):

$$MAE = \frac{\sum_{i=1}^{n}|yi - y'i|}{n} \qquad \ldots\ldots (6)$$

Where n: number of data rows, yi: actual target value of record i, y'i: predicted target value of record i.

- Mean Squared Error (MSE):

$$MSE = \frac{\sum_{i=1}^{n}(yi-y'i)^2}{n} \qquad \ldots\ldots (7)$$

The parameters as explained above.

- Relative Absolute Error (RAE):

$$RAE = \frac{\sum_{i=1}^{n}|yi - y'|}{\sum_{i=1}^{n}|yi - y''|} \qquad \ldots\ldots (8)$$

Where y'': average of original target values, other parameters as explained in equation (2.6) above.

- Root Mean Squared Error with Log (RMSElog):

$$RMSE\ log = \sqrt{\frac{\sum_{i=1}^{n}(\log(yi+1)-\log(y'i+1))^2}{n}} \qquad \dots (9)$$

The parameters as explained in equation (2.6) above, log: natural log.

Cross Validation

Dividing an available dataset between testing and training processes may lead to an unreliable evaluation for the model [2, 3]. The problem here is that the selected part of data could be non-representative, for all data. The solution is to use **cross-validation**. It repeats the whole process (i.e. training and testing) many times with different samples of data records. In three cross-validations, prediction models would be repeated three times. In each trial, one-third of the data is selected for testing and the remainder is used for training. The error of all iterations is averaged to get an overall error rate [23, 22].

Designing a Boosted Prediction Model for Big Data of HealthCare

Introduction of the Proposed System

The predictive analysis involves a variety of statistical, modeling, data mining, and machine learning techniques to study and analyze historical data, thereby giving data the ability to make predictions about future information. Predictive analysis of healthcare data sets involves the analysis of historical medical records of patients to make an approximate idea about their future health status. Prediction of future hospitalization could reduce unnecessary costs, which would be considered one of the advantages in healthcare financial aid management. There are two factors which control hospitalization costs: (i) the number of patients who entered the hospital, (ii) how long each patient stays.

Many problems could be mentioned here, such as inaccurate estimation of the period from the medical staff. This increases the waste of resources and the high percentage of historical records which have a zero value of days in the hospital. This means that the data has an imbalance distribution making the predictive process more complicated. The primary challenge is to build an algorithm of a prediction model aiming to solve the problems above with precise prediction in less time.

The goal of this prediction model is to predict precisely how long each patient will be admitted to the hospital depending on his medical records. It supports the medical staff to make the correct decision about the real need of the patient and reduce the costs. Therefore, this process can be used to enhance particular services and government health institutions. In this work, a Boosted Prediction Model for Big Data of HealthCare (BPM-BDHC) is presented to solve healthcare problem by replacing the split criteria of a boosted regression tree with correlation measures. Afterward, the resulting model is evaluated by many error measures based on real and complex techniques to get the lowest prediction error in less time. BPM-BDHC has three major stages: *(i)* preprocessing step (i.e., dataset is made more suitable for the next stage, *(ii)* building a prediction model using improved boosted regression, *(iii)* evaluating the result of the previous stage. The steps are explained in Figure (6).

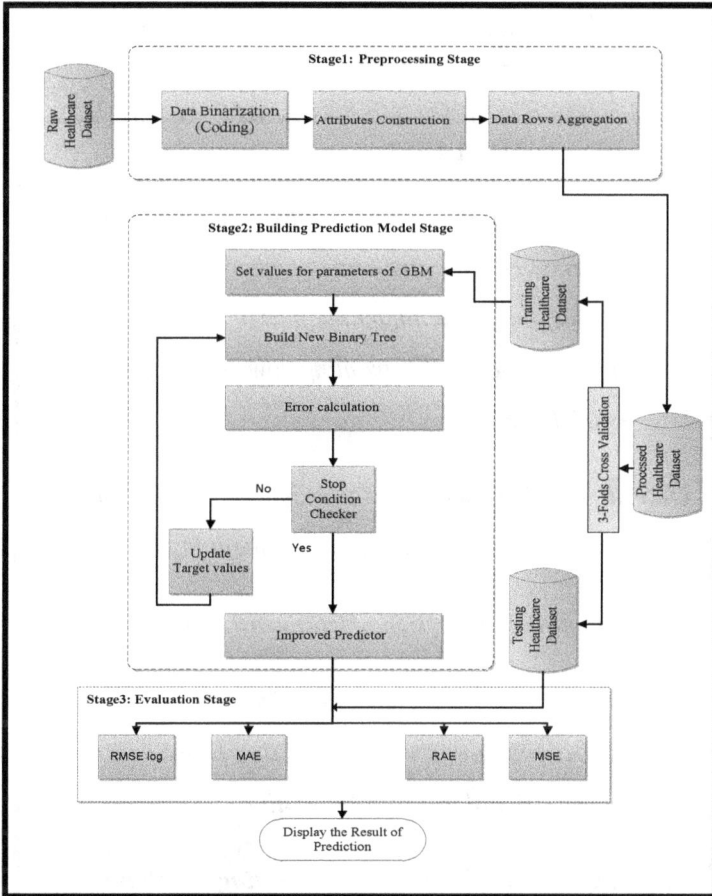

Figure (6) Block Diagram of the main steps of BPM-BDH

In this section, the outcome of replacing traditional splitting criteria by correlation based rules is analyzed and explained. BPM-BDHC includes three stages as described in Algorithm 1;

the first step is data set preprocessing which includes data binarization, attributes construction and data row aggregation. The second phase involves building a prediction model including parameter detection and formation of regression trees. The final stage is the evaluation of the results based on the gained test data and the use of the cross-validation concept.

Algorithm 1: MPM-HCP

Input: *D: Healthcare Dataset, Tmax: maximum number of trees Tnmax: maximum number of terminal nodes, Smin: minimum number of samples in a terminal node, RC: training row count, Shr: shrinkage.*

Output: *Tmodel: regression trees model, Number of days in a hospital for each patient in next year.*

Step 1: *Call Pre-Processing Healthcare Dataset (D).*

Step 2: *Split Processed Healthcare Dataset into Training dataset Tr and Testing Dataset Ts using 3-Folds Cross Validation.*

Step 3: *Call **Modified Gradient Boosting Machine Building** (Tr, Tmax, Tnmax, Smin, RC, Shr) and retrieve Tmodel.*

Step 4: *Call Testing Trees model (Tmodel, Ts, Shr)*

End

Preprocessing Stage

Real datasets, like healthcare datasets, may have some inappropriate structures that need preprocessing before utilizing them. As we explained in (2.2), preprocessing may deal with incomplete data or data with the unsuitable structure for mining. Algorithm (3.2) summarizes the preprocessing steps.

Algorithm 2: Preprocessing Healthcare Dataset
Input: *D: Raw Healthcare Dataset.*

Output: *Processed Healthcare Dataset.*

***Step 1**: For each categorical attribute (ca) in D, apply data binarization by:*

    1.2   *For each category value in ca:*

        1.2.1   *Create new empty attribute (na).*

1.2.2 *Set the value of new attribute (na) by:*

        o   *For each data row Rc in D*

            If Rc[ca] = na then

            Rc[na]=1

            else

            Rx[na]=0

            end if

***Step 2**: From each numerical attribute na in D, construct new attributes from*

    mathematical operations by:

    2.1   *Find all data rows that related to the same patient identifier*

        and same year.

    2.2   *Find Min, Max, Average and Standard Deviation for those data*

        rows.

***Step 3**: For all data rows in D resulted from step1, apply data aggregation by:*

    3.1   *Find all data rows that related to the same patient identifier*

        and same year.

    3.2   *Aggregate those data rows to be a single row.*

***Step 4**: For all tables in D, create one table from the combination between them.*

End

Next, BPM-BDHC uses a correlation measure to build a prediction model, and this requires transforming categorical attributes of a healthcare dataset into numerical attributes.

Binarization, which was explained above, includes expanding original attributes by converting each absolute value to the new attribute. Figure (7) illustrates the idea of binarization, in this example, of healthcare data.

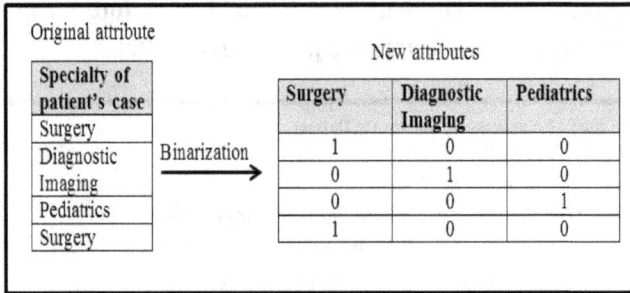

Figure (7) Data Binarization for Example of Healthcare

Attributes Construction

A new set of attributes is constructed to extend the original numerical attributes for reducing the error of prediction. This is performed by calculating statistical information such as maximum value, minimum value, the average and the standard deviation for all claims for the same patient in a particular year. Figure (8) illustrates attribute construction with data.

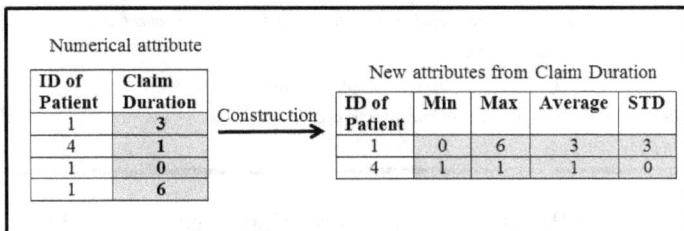

Figure (8) Attribute Construction for Example of Healthcare Dataset

Data Rows Aggregation

As derived from step A (i.e. claims on healthcare), all data rows representing each patient in a particular year are aggregated into a single row. The goal of predicting a healthcare dataset is to anticipate the number of days that will be spent in the hospital by the patient. This requires aggregating separate information from all claims together in the training data.

Tables Combination

All tables of a healthcare dataset are combined into one table to include all personal, hospitalization, drugs and lab information together. The reason for this step is to reduce the complexity of the training process and hence increase the efficiency of the tree building process. This is more efficient than dealing with partitioned tables.

Building Prediction Model Stage

This stage could be considered the core of our MPM-HC approach. It starts with the detection of some critical parameters which are needed in boosted regression. With the utilization of the iteration process, the binary regression tree is built to reduce the error of the previous one until the stop condition is satisfied. The BPM-BDHC model *replaces the criteria of choosing the best splitter attribute inside the binary tree procedure, with a more efficient standard.* The steps of this stage are clarified in Algorithm 3.

Algorithm 3: Modified Gradient Boosting Machine

Input: Tr: training data, Tmax: maximum number of trees, Sk: learning   rate,
Tnmax: number of terminal nodes, Smin: number of data records in a
terminal node, N: number of data records in D, y: index of the target.

Output: Prediction Hospitalization Boosting Model.

Set: Fx: an array of predicted values of training data rows, rc: rows counter, Tc: trees
counter. Org_target: an array of the original target of Tr.

Step1: Find the initial prediction for all data records in Tr by:

      1.1   *Calculate the mean of target values Mean(Y).*

      1.2   *While rc< N*

          1.2.1   *Fx [0, rc] = Mean(Y)*

          1.2.2   *Org_target[rc] = Tr[y,rc]*

          1.2.3   *Increase row counter: rc = rc + 1*

          End While

Step2: While Tc<=Tmax, build boosted model by:

      2.1   *rc = 0*

      2.2   *While rc< N, Update target values of Tr by:*

          2.2.1   *Residual [rc] = Org_target [rc] – Fx[Tc-1,rc]*

          2.2.2   *Tr[y,rc] = Residual [rc]*

          2.2.3   *Increase row counter: rc = rc + 1*

          End While

      2.3   *Call Improved* **Regression Tree Building** *(Tr, Tnmax, Smin)
and retrieve T.*

Step3:   *For each terminal node Tn in T by:*

      3.1   *While rc_tn< number of data rows in Tn, update prediction
value by:*

          3.1.1   *Fx [Tc,rc_tn] = Fx [Tc-1, rc_tn] + (Sk
×Tn.predicated_value)*

          3.1.2   *Increase counter of data rows in Tn: rc_tn = rc_tn + 1
End While*

      3.2   *Increase trees counter: Tc = Tc + 1*

      End While

Step3: Return boosted trees model Tmodel with array of prediction values Fx.

End

Parameters Gathering

The first step to building a robust prediction model is choosing the parameters of the algorithm carefully. Gradient Boosted Regression has four main parameters which should be considered: (*i*) maximum number of trees in the model that control the execution of the algorithm, (*ii*) maximum number of terminal nodes in every single binary tree which control the number of rules, (*iii*) minimum number of samples in each terminal node that effect on the decision in that node, (*iv)* shrinkage, which represents the learning rate in the training process.

Build Binary Regression Tree

When a current number of trees in the model is still less than a maximum number of trees, a new binary tree is created. Each tree works with the same set of attributes. However, the values of the target of prediction would be updated by the prediction of previous trees, and they must be minimized. The steps of building a new regression tree are explained in Algorithms (4).

This algorithm consists of a recursion procedure to create the regression tree. It can be described by the following four points:

1. Best Splitter Attribute Choice

    The most important step in constructing a binary regression tree is to choose a splitter attribute which has more relations with the target, and this is according to specific criteria. The binary tree for a classification problem depends on information gain, gain ratio and Gini index for choosing the best attribute. On the other hand, a regression problem uses SSE and Standard Deviation.

BPM-BDHC developed this important step by using the correlation between each attribute and the target as an indicator of attribute quality for splitting operations. This is performed by applying relationship quality selection in pre-processing. The equation (4) of Pearson's correlation, is employed in this step to evaluate the advantages of each attribute, as shown in Figure (4).

2. Best Split Point Choosing

BPM-BDHC uses a straightforward and efficient method to choose the best split point depending on the fraction of squared summation of target values dividing by the number of data rows for both right and left child node. The equation (2.4) explains the calculation of split error.

3. Data Splitting

In this step, current data is divided into two parts according to the condition of the best split point, which was detected above. The data rows whose splitter attribute has a value less than that of the split point, become the information of a new left child node. While the data rows consisting of a value equal or larger than that of the split point, become the input of a new right child node.

4. Prediction Value in Terminal Nodes

Each path from the root of a tree to a terminal node has rules which consist of many conditions. The control should lead to predicting a particular value for the target of prediction. Mostly, the terminal node has more than one data row, and each one possesses a target value. The prediction value of a particular node simply could be the mean of all values of the data rows in the node itself.

Training Error Calculation

Gradient Boosted regression depends on the concept of minimizing the gradient of error. The error of every data row is calculated by subtracting the original target value of that data row from its predicted value. The average training error for a specific tree could be found by dividing the summation of the error of all the data rows by their number.

Target Values Updating

The value of the target in each data row is updated by adding the prediction of the new tree and multiplying it by the shrinkage rate. This step is considered the most important in an improved GBM and it differentiates this technique from other prediction methods. The example in Figure (4) clarifies this step. It could be explained as the search for the best value which is closest to all the target values.

Evaluation Model Stage

In this stage, all tree models built by a Gradient Boosted Regression are evaluated based on a test dataset which has never been seen before. RMSE Log, MSE, RAE and MAE measurements are used in this step to assess the prediction error of every tree and then find the total error through the combination of the whole tree model. The concept of Cross Validation is used to get a more reliable evaluation. The process of training and testing is repeated three times (i.e. three folds.).

Algorithm 4: Improved Regression Tree Building

Input: *Tr Healthcare Training Dataset, Tnmax: maximum number of terminal nodes, Smin: minimum number of samples in a terminal node, N: Training rows count.*

Output: *RegressionBinary Tree.*

Set: *Rcc: current rows count, Tnc: Terminal nodes counter, rc_tn: counter of data rows in a terminal node, Tnc: counter of trees.*

Step 1: *If (Rcc>= Smin) then, do the following:*

1.1	*Create Terminal Leaf Node Lf*
1.2	*Lf.predicated_value = mean (Trt)*
1.3	*Increase the number of terminal nodes*
	$Tnc = Tnc + 1$
1.4	*Return Lf*

Else

1.5	*Create internal node INode*
1.6	*Choose best splitter attribute by:*

      1.6.1   *For each attribute xj in attributes set, find correlation wj between xj and target according to the equation (2.4).*

      1.6.2   *Choose best quality (bs_attr) with highest correlation value as a splitter for node INode.*

      1.6.3   *Choose splitting point sp on bs_attr which minimizes error according to the equation (3.2).*

      1.6.4

1.7	*Split Data Tr based on sp into two parts: TrL, TrR.*
1.8	*INode. attribute _splitter =bs_attr.*
1.9	*INode.splite_point = sp .*
1.10	*INode.Left_child = Call Improved Regression Tree building (TrL, Tnmax, Smin, Rcc)*
1.11	*INode.Right_child = Call Improved Regression Tree building (TrR, Tnmax, Smin, Rcc)*

End if

Step2: *return binary regression tree T.*

End

Implementation of BPM-BDHC with

The efficiency of BPM-BDHC described in the previous chapter has been tested with different parameter values, and the results will be reviewed in this chapter. A real and huge dataset has been used as an implementation case study to discover the behavior of that model. The experimental result of each stage of BPM-BDHC is shown and explained.

The insufficiency of real and reliable healthcare data in Iraqi health government institutions led us to utilize a different dataset in this implementation. It presents extensive and trusted healthcare data, and is considered a demonstration of the ability of BPM-BDHC to handle challenging and complicated datasets and it is taken from URL[1].

Description of a Healthcare dataset

Heritage Provider Network (HPN) provides a healthcare dataset for researchers and data miners aiming to reduce the costs by predicting future hospitalization of patients. It contains data from more than (113000) patients in eight tables linked by a patient identifier. This is described in Table (1).

Each patient has one or more claims a year, the data of application includes information about the condition causing hospitalization and the medical procedure required for patient treatment. Attributes of a claims table are explained in Table (2).

---

[1]https://www.heritagehealthprize.com

Table (1) Description of a Healthcare dataset

Table Name	Description	Number of Attributes
Claims	The main table for healthcare datasets and contains information about the medical case of the patient.	14
Members	Includes personal information such as age.	3
DaysInHospital_Y2	Provides the number of days spent by the patient in the second year; it used in the training stage.	3
DaysInHospital_Y3	Contains the number of days spent by the patient in the third year; it is employed in the preparation stage.	3
DrugCount	Drugs consumed by the patient.	4
LabCount	Number of lab tests consumed by the patient.	4
PrimaryConditio_nGroup	Describes primary condition group coding.	2
LookupProcedur_eGroup	Describes medical procedure group coding.	2

Table (2) Description of Claims table

Attribute Name	Description	Number of Categories
MemberID	Patient identifier, which links all tables.	-
Year	The year of claim Y1; Y2; Y3.	3
Specialty	General specialty of patient's condition.	13

PlaceSvc	General place of healthcare service.	9
PayDelay	The duration between the date of medical service and date of payment (in days).	163
LengthOfStay	Days of hospitalization (i.e. delay between discharge date and admission date).	10
DSFS	Days since the first claim.	13
PrimaryCondit_io nGroup	Code of medical diagnostics described in the primary condition group table.	46
CharlsonIndex	A value of affecting diseases.	4
ProcedureGro_up	Code of procedure diagnostics described in the procedure group table.	18
SupLOS	The binary value of suppression of claim.	2

The values of days in the hospital are between (0 and 15). A duration of more than 15 days is rounded up to 15 in this dataset. Grouping target values in specific ranges could make the task as a multi-class classification complicated. Handling of the continuous value turns the prediction function into a regression problem. BPM-BDHC treats this dataset as a regression model to get (1) more precise (2) reliable results, which can (i) reduce the total error of prediction (ii) prevent wasting unnecessary costs. The distribution of days in hospital values is shown in Figure (9).

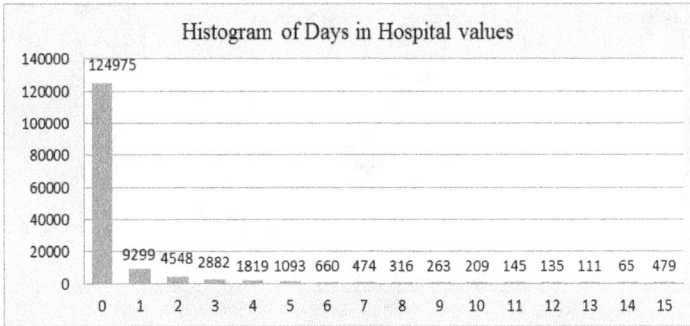

Figure (9) Histogram of Days in Hospital values

Applying BPM-BDHC on a HPN Healthcare dataset

Detailed and comparative explanations of using BPM-BDHC on a HPN healthcare dataset is introduced in this section.

Preprocessing stage utilizing

The preprocessing step of BPM-BDHC is used on HPN datasets. The reason for applying these measures is due to the presence of many tables with different types of data sets, which make the training process more complicated. The steps of applying preprocessing on HPN datasets produced (128) new attributes from (13) primary attributes. This could be explained as follows:

A. Data Binarization

Transforming attributes to binary form are applied to HPN datasets. This step is performed on categorical attributes; Age,

Gender, Primary Condition Group, Specialty, Procedure Group and Place Svc attributes. It yields (97) new attributes.

B. Constructing New Attributes

BPM-BDHC utilizes some mathematical operations on HPN datasets to discover further hidden information. This step includes calculating and creating new attributes for some claims of the patient in a particular year. This depends on the number of specialties, the number of Primary Conditions and the number of Medical procedures. Furthermore, it includes calculating the minimum, maximum, and average and STD values for Pay Delay, Length of Stay, DSFS, Charlson Index, Drugs, and Labs. It constructs (32) new attributes for those values.

C. Data Rows Aggregation

In HPN datasets, the number of data rows in the claims table would be reduced by more than one million to 147473 records by the process of aggregating. The major questions here, why are we aggregating the claims of patients? And what is the benefit of this step? Is this approach making the complexity of the procedure more or less? In a period of a year, all requests made for a particular patient are aggregated, which make the process of choosing the best attribute as a splitter in the building a tree step more efficient. This aggregation reduces the overall complexity of the prediction problem, and this occurs through the knowledge of the HPN datasets.

Therefore, the benefit of the preprocessing stage is clear. The difference between the training error before and after applying steps of preprocessing is distinguishable. The gradient boosting machine (GBM) had a 0.668 error measure when used on HPN datasets without preprocessing. The error was reduced after 80 iterations to 0.49. While, preprocessed datasets produced a mistake starting from 0.515, and ending with 0.48, according to the RMSE log measure.

Implementation of Prediction Model Stage

Before building the prediction model, we need to combine all tables of HPN datasets explained above into one table that contains (128) attributes, including the target of speculation and (147473) the data rows. This dataset is considered as an input to the training process and builds a prediction model. An HPN dataset is used to create a boosted set of regression trees in which the training error is optimized. The dataset has been split according to the cross-validation concept. It includes three folds and within each fold, the HPN dataset is divided into two parts; 2/3 for the training stage (98326 records) and 1/3 for the testing step (49157 records).

A. Applying Parameter Values Selection

The selection of the parameter roles that indicates the behavior of the training process affects the result of this procedure. Multiple values of improved Gradient Boosted Machine parameters have been used with HPN datasets aiming to target their optimal values. The *shrinkage parameter* is tested with a range of the values (0.1 – 0.001), and the best result was obtained when using the range (0.005- 0.008) on the other parameters.

The other parameter that should be selected for GBM is the maximum number of terminal nodes which relates to the complexity of the trees and the number of rules. The standard depth of a regression tree in GBM is between (2 and 4), which means the maximum number of terminal nodes is in the range of (4-16) nodes. In BPM-BDHC, the depth of three, four and five levels is tested, and results are obtained using five levels. This means that the maximum number of terminal nodes within an HPN dataset is (32) nodes, which is considered medium complexity for a dataset with (128) attributes.

The third parameter in GBM is the minimum number of records in the terminal nodes. It means that if the number of data rows in the node is equal to or less than this parameter, the node would be the final one and the split operation would stop. This parameter was set to (20) records with an HPN dataset.

The number of trees is the most important parameter in this process which needs to be chosen carefully. With an HPN dataset, the range of (80 – 160) tree is used. The effect of increasing the number of trees could be seen in Figure (10).

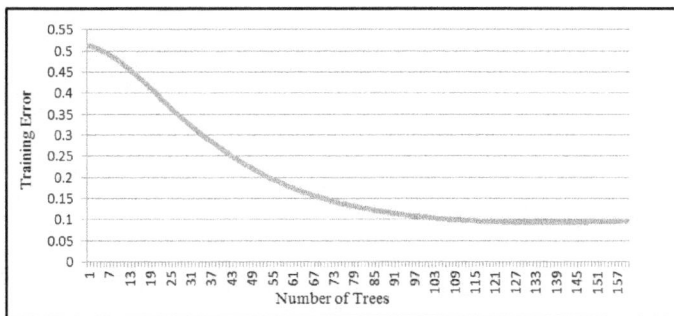

Figure (10) Training Error of BPM-BDHC in model of 160 trees.

In the original GBM, the training error is reduced and reaches to the lowest error of (0.476) after 97 iterations of training, then it is increased again. The training error of BPM-BDHC was gradually reduced until it reached its lowest value of (0.093) after 136 iterations. This represents a definite improvement compared with the original GBM.

B. Implementation of Improved GBM

The parameters which have been set in the previous section will be used to build a binary regression tree. In the first iteration, the initial guess of the target attribute is used rather than building a tree. It is the average of all target values. An HPN dataset has more than 89% of its goal values set as a zero value. For this reason, the initial guess in BPM-BDHC prefers to be zero.

The main improvement of BPM-BDHC involves replacing the traditional splitting criteria of the regression tree by using the correlation which leads to enhancing the training error. Two observations could be seen after building (160) regression trees in both the original and improved GBM:

- **First:** The training error of standard deviation is based on GBM and SSE. GBM starts to increase after the creation of (100) trees, in both cases. Also, there was a matching in the results of the two measures.

- **Second:** BPM-BDHC shows a more stable behavior and less training error with both original GBM methods. Furthermore, the gradient of mistakes doesn't increase even if (160) trees were built.

Both observations are shown in Figure (11).

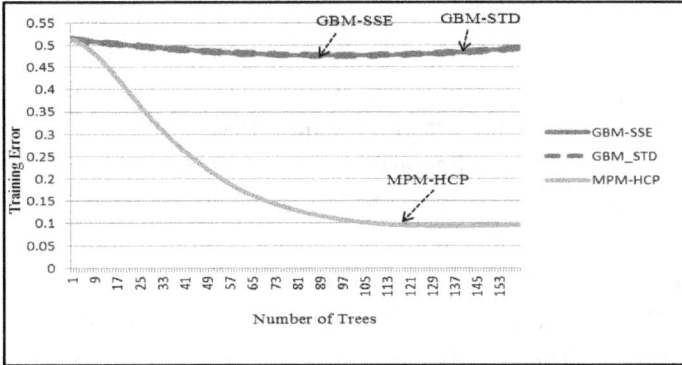

Figure (11) Comparison Training Errors of GBM and BPM-
BDHC

Evaluating the Prediction Model

Reliable evaluation of data mining techniques should be based on test data which has never been seen before during the training stage. According to cross-validation, an HPN dataset is inserted in each fold and split into two parts in which the experimental data is 1/3 of the original data, and the rest is used for training purposes. This means that the number of test data rows is (49157) records. The training stage is terminated with the building of (160) regression trees that need to be evaluated, and a three folds cross-validation technique is applied.

Evaluating based on Error:

There are many possible methods to test the final model (i) taking the last tree for testing purposes only (ii) randomly choosing trees to contribute to the evaluation process (iii) utilizing all trees built in the training process for evaluation of the model and control contribution of each tree by the learning rate parameter. The first method depends on the last tree which has an accumulative model for all the previous trees. In this approach, the original GBM based on SSE and STD has a (0.586) average testing error in the three folds and RMSE log measure.　The BPM-BDHC testing error is (0.48) which represents the best result in the first method. Table 3 shows the comparison of this approach using four error measures.

Table (3) Comparison of Testing Error of GBM and BPM-BDHC in Last Tree

	Name of Method	MAE	RAE	MSE	RMSE Log
Fold 1	GBM-SSE	0.903645	1.167224	2.460890	0.586971
	GBM-STD	0.903645	1.167224	2.460890	0.586971
	BPM-BDHC	0.519348	0.675684	2.591641	0.480892
Fold 2	GBM-SSE	0.904521	1.192644	2.481206	0.588235
	GBM-STD	0.904521	1.192644	2.481206	0.588235
	BPM-BDHC	0.520267	0.677308	2.554392	0.481644
Fold 3	GBM-SSE	0.892517	1.207927	2.388095	0.584159
	GBM-STD	0.892517	1.207927	2.388095	0.584159
	BPM-BDHC	0.520409	0.677494	2.554360	0.481694

Discussion

In this chapter, we dealt with real data which was characterized by its magnitude and high complexity. GBM depends on the binary regression tree in the construction of the model, and that leads to three essential obstacles, namely:

- Main Memory Consuming:

GBM uses recursion in building a BRT since it requires the presence of all the data in the memory during training time. It consumes a significant storage source in case of large data. In our proposed model, the problem has been solved by using various methods of memory optimization (such as using pointers and local variables that are deleted from the memory after the end of the task).

- Execution Time:

A large number of data rows and the number of attributes in the healthcare dataset needs a long time to be processed using typical prediction techniques. The training process includes many splitting and comparing steps for tree building, and these steps are repeated many times to create sequences of BRT. The required time to execute all these steps is considered a problem and we aimed to overcome it in our proposed model by using the correlation splitting criteria which reduces the comparing steps.

- Parameters Detection

Four basic parameters must be carefully selected in the improved GBM. These are (the minimum number of samples that can be contained in the terminal node, the maximum number of terminal nodes, the maximum number of trees allowed in the model and the learning coefficient). Both the first and the second parameters affect the building of the tree and decision-making, while the third and fourth parameters are related to the process of reducing the error rate in the construction of the integrated model.

The process of choosing values for such parameters is critical and affects the results of training. Therefore, this step must be carried out very carefully. In our proposed model, we used the principle of trial and error in choosing parameter values. Hence, we did not employ any particular strategy. Mostly techniques of sampling are used as a part of preprocessing significant data, and that leads to creating unrealistic results especially when it comes to healthcare data. In our proposed model, BPM-BDHC, we did not use any sampling technique. Nevertheless, our method provides precise, reliable and realistic results.

Summary

Healthcare datasets are complicated and pose various problems which are solved by using MPM-HCP. The correlation between the attributes and the target of prediction are introduced and utilized as the new splitting criteria to choose the best quality during the building of a binary regression tree. BPM-BDHC has significant behavior regarding the prediction error and execution time. In comparison with the standard gradient boosting trees, the proposed system gave a clarified improvement in the training process. The training error was close to zero; the testing error was better than SSE and the standard deviation was based on GBM. Furthermore, the training time is reduced to over 85%

which gives BPM-BDHC more scalability to deal with a broad and complex healthcare dataset.

REFERENCES

[1] Raymond L. Goldensteel, and Karen Goldensteel, "U.S. Healthcare system", 7th Edition, ISBN: 978–0-8261–0930-9, Springer Publishing, 2013.

[2] Pang-Ning Tan, Michael Steinbach, and Vipin Kumar, "Introduction to Data Mining", ISBN-13: 978-0321321367, Addison-Wesley, 2005.

[3] Jiawei Han and MichelineKamber, "Data Mining: Concepts and Techniques", 2nd Edition, ISBN-10: 1-55860-901-6, Elsevier, 2006.

[4] DivyaTomar and SonaliAgarwal, "A survey on Data Mining approaches for Healthcare," International Journal of Bio-Science and Bio-Technology 5.5 241-266, 2013.

[5] Jared Dean, "Big Data, Data Mining, and Machine Learning: Value Creation for Business Leaders and Practitioners", ISBN-13: 978-1118618042, Wiley Publishing, 2014.

[6] Jiawei Han and MichelineKamber, "Data Mining: Concepts and Techniques", 3th Edition, ISBN 978-0-12-381479-1, Elsevier, 2013.

[7] Trevor Hastie, Robert Tibshirani , and Jerome Friedman, "The Elements of Statistical Learning", 2nd Edition, ISBN 0-387-84857-6, New York: Springer, pp. 337–384, 2009.

[8]   Barry deVille, "Decision Trees for Business Intelligence and Data Mining: Using SAS Enterprise Miner", ISBN-13: 978-1590475676, SAS Publishing, 2006.

[9]   StéphaneTufféry, "Data Mining and Statistics for Decision Making", First Edition, ISBN: 978-0-470-68829-8, John Wiley & Sons, 2011.

[10] Jerome H. Friedman, "Greedy function approximation: a gradient boosting machine", Annals of Statistics: 1189-1232., 2001.

[11]   J. Elith, J. R. Leathwick, and T. Hastie, "A working guide to boosted regression trees", Journal of Animal Ecology, vol (77), p.p 802–813, 2008.

[12]   Dimitris Bertsimas, Margrét V. Bjarnadóttir, Michael A. Kane, J. Christian Kryder, RudraPandey, SantoshVempala, Grant Wang, "Algorithmic Prediction of Health-Care Costs", OPERATIONS RESEARCH, ISSN 0030-364X, Vol. 56, No. 6, November–December 2008, pp. 1382–1392, 2008.

[13]   Xiang Peng, Wentao Wu, and JiaXu, "Leveraging Machine Learning in Improving Healthcare", Association for the Advancement of Artificial Intelligence, 2011.

[14] Rashedur M. and Fazle Rabbi, "Using and comparing different decision tree classification techniques for mining ICDDR, B Hospital Surveillance data", Expert Systems with Applications vol (38), p.p 11421–11436, 2011.

[15] HadiYazdi and NimaMoghaddami, "Multi Branch Decision Tree: A New Splitting Criterion", International

Journal of Advanced Science and Technology, Vol. 45, August 2012.

[16] Jufen Zhang, Kevin M. Goode, Alan Rigby, Aggie H.M.M. Balk and John G. Cleland, "Identifying patients at risk of death or hospitalization due to worsening heart failure using decision tree analysis: Evidence from the Trans-European Network-Home-Care Management System (TEN-HMS) Study", International Journal of Cardiology, vol (163), p.p 149-156, Elsevier, 2013.

[17] Nannan He, "Data Mining for Improving Health-Care Resource Deployment", Master thesis, University of California Santa Cruz, 2014.

[18] NittayaKerdprasop, KittisakKerdprasop, "Moving Data Mining Tools toward a Business Intelligence System", World Academy of Science, Engineering and Technology 25, 2007.

[19] Sanford Weisberg, "Applied Linear Regression", 3th Edition, ISBN-13: 978-0471663799, A John Wiley & Sons, Inc. Publication, 2005.

[20] Ian H. Witten and Eibe Frank, "Data Mining: Practical Machine Learning Tools and Techniques", 2nd Edition, ISBN-13: 978-0120884070, Morgan Kaufmann, 2005.

[21] Robert Nisbet, Gary Miner, and John Elder,"Handbook of Statistical Analysis and Data Mining Applications", ISBN-13: 978-0123747655, Academic Press, 2009.

[22] Max Kuhn and Kjell Johnson, "Applied Predictive Modeling", ISBN 978-1-4614-6849-3, Springer, 2013.

[23] Ian H. Witten, Eibe Frank, and Mark A. Hall, "Data Mining: Practical Machine Learning Tools and Technique", 3nd Edition, ISBN-13: 978-0123748560, Morgan Kaufmann, 2011.

[24] Thomas Dietz and Linda Kalof,"Introduction to Social Statistics: The Logic of Statistical Reasoning", ISBN-13: 978-1405169028, Wiley-Blackwell, 2009.

[25] Anthony Graziano and Michael Raulin, "Research Methods: A Process of Inquiry", ISBN-13: 978-0205907694, 8th Edition, Pearson Publishing, 2012.

[26] Thomas Cleff, "Exploratory Data Analysis in Business and Economics: An Introduction Using SPSS, Stata, and Excel", Print ISBN-13: 978-3329015163, Springer; 2014.

[27] Luis Torgo, "Inductive Learning of Tree-based Regression Models", Ph.D. Thesis, University of Porto, Portugal, 1999.

[28] Zhi-Hua Zhou, "Ensemble Methods: Foundations and Algorithms", ISBN-13: 978-1439830032, Chapman and Hall/CRC, 2012.

[29] RobiPolikar, "Ensemble based systems in decision making", Circuits and Systems Magazine, IEEE (Vol: 6, Issue: 3) p.p 21 – 45, 2006.

[30] LiorRokach and OdedMaimon, "Data Mining with Decision Trees: Theory and Application", ISBN-13: 978-9812771711, World Scientific Publishing Company, 2008.

[31] GiovanniSeni and John F. Elder, "Ensemble Methods in Data Mining: Improving Accuracy Through Combining

Predictions", ISBN-13: 978-1608452842, Morgan and Claypool Publishers, 2010.

[32] RomanTimofeev, "Classification and Regression Trees (CART) Theory and Applications", Master Thesis, Center of Applied Statistics and Economics, Humboldt University, Berlin, 2004.

[33] Yoav Freund, "An Adaptive Version of the Boost by Majority Algorithm", Machine Learning, 43, 293–328, Kluwer Academic Publishers, 2001.

CHAPTER FIVE

IoT SECURITY

Abdulsahib Albehadili[1] and Dr. Ahmad Javaid[2]

[1]Department of Electrical Engineering and Computer Science, University of Toledo, OH, USA

[2]Department of Electrical Engineering and Computer Science, University of Toledo, OH, USA

ABSTRACT

The Internet of Things (IoT) is the vision of the future internet where physical objects (like sensing devices, vehicles, buildings, etc.) collect and exchange data. The IoT allows objects to be sensed and controlled remotely across existing network infrastructure. It provides direct integration of the physical world into computer based systems, resulting in improved efficiency and accuracy. Conventionally, security has been an enabling factor in the communications world. Similar to other technological applications handling sensitive information, it is necessary for IoT applications to be developed and deployed with highly secure mechanisms to protect data and services provided by these technologies. Existing security approaches need to be improved to make them adapt to the requirements of the IoT platforms, platforms that lack resources and computational power.

This chapter presents existing security approaches being used for IoT, together with the weaknesses they inherit. Since security of IoT communications could be addressed in the context of the communication protocol itself, we focus on existing protocols and mechanisms used to secure communications in the IoT. We also analyze how these protocols ensure fundamental security requirements and protect communications on the IoT, together with the open challenges and strategies for future research work in the area.

Keywords: Internet of things (IoT), IEEE 802.15.4, CoAP, 6LoWPAN, DTLS, RPL, security.

INTRODUCTION

The development stage of IoT is to integrate networks of devices, having unique characteristics and deployment contexts that would operate without human intervention for long periods of time. Various threats to the security of IoT create an array of new problems concerning breach of privacy and information security. Those threats could affect *frontend* sensors and equipment, *networks*, as well as *backend* of IoT systems.

In a typical Wireless Sensor Network (WSN), data is received via the distributed sensors and then transmitted using modules of Machine-to-Machine (M2M) devices, thus achieving networking services of multiple sensors. If an intruder can gain access to these sensors, damage or an illegal action is possible; such as unauthorized access to data, denial of service attacks, or privacy analysis of M2M.

On the other hand, the *network* plays a vital role in providing the interconnection capabilities to the sensors, as well as the reliability of a rich quality of service (QoS). Since a large number of devices in the IoT rely on the network to send their data, a network connection compromise can easily result in a denial of service attack. In the same context, the backend is the core of any IoT system and thus has a vital role in the various sensing applications. Accordingly, critical security measures are required, as these systems form the gateway and the middleware of the IoT. They handle sensor data in real time, or sometimes in pseudo real-time, to increase service intelligence and QoS.

Furthermore, to gain the customer trust, an IoT system should be built to conform to major security standards in the areas including, but not limited, to access control, communication layer security, user authentication and authorization, privacy protection, data integrity & confidentiality, and data availability at any time [1].

Nonetheless, privacy of IoT devices is one of the critical challenges to be addressed. Privacy could be seen as the right of a node to act on its behalf, which means it is the degree to which the node determines its desire to share data about itself with others and also to interact with its environment. In IoT, the environment is sensed by connected devices. Then, these sensing devices share their observations with core servers where the application logic is carried out. Thus, privacy should be protected at all levels: in the node itself, in the storage, as well as during communication.

An unauthorized reconfiguration of the firmware or the hardware of the devices within the network may leak out sensitive data. For instance, a surveillance camera could be reconfigured to redirect data to an intruder instead of the legitimate server. Thus, strict privacy measures should be considered for devices that gather sensitive data. There are many issues that need to be addressed to provide privacy in those devices; such as the location privacy, non-identifiability (protecting the identification of the exact nature of the device), and personal information privacy (in case of device theft or loss, and resilience to side channel attacks). As an example, the location privacy could be realized using *multi-routing random walk algorithm* in WSN, and the *quick response codes* (QR) technique could be deployed for display privacy and *personal identifiable information* (PII) in the case of device loss.

On the other hand, privacy during communication is ensured using various encryption techniques, as it is the common approach that enables data confidentiality during data transmission. On certain occasions, encryption adds extra data bits to packets headers which provides a way for tracing back: for instance, sequence numbers, *IPsec - security parameter index*, etc. This critical data could be compromised by a cryptanalyst who could use it to analyze the traffic flow. To overcome this problem, secure communication protocols could be a suitable approach.

For storage privacy, some measures need to be considered for protecting data. For instance, only the least possible amount of data should be stored. In inevitable cases, only personal data is retained. To reveal the real identity corresponding to specific stored data, *pseudonymization and anonymization* may be used. It is also possible to disclose only statistical data, like sum, average, and count, rather than specific records. An appropriate technique called *differential privacy* could assure the independence of an output and a particular record by noise addition [1].

So far, we have stated the security and privacy requirements that should be met by an IoT system. This would give the reader a high-level overview of the enabling factors for IoT technology. Next, we present and analyze the existing security technologies for IoT communications, as well as the ones that are currently being developed.

SECURITY IN CURRENT IOT COMMUNIC-ATIONS TECHNOLOGIES

Connectivity of IoT devices will be mostly based on IP communication protocols. These protocols are being developed in line with the constraints of the sensing platforms likely to be deployed by IoT applications. Such development integrates a protocol stack able to provide power-efficient & reliable internet connectivity.

Security has always been an enabling factor in the communications world. This makes it necessary for IoT applications to be developed and deployed with highly secure mechanisms to protect services provided by these technologies. Furthermore, existing protocols must be analyzed to secure communications in the IoT platforms. In this section, we present existing security approaches, together with the weaknesses they inherit, and the open challenges that need to be addressed.

IoT paradigm integrates WSN, M2M communications and Low Power Wireless Personal Area Networks (LoWPAN), or technologies such as Radio-Frequency Identification (RFID). These technologies, in their nature, impose restrictions regarding the resources availability in the sensing platforms: like memory, and computing power. However, trends to design and adopt communications and security technologies optimized for constrained sensing platforms is reflected by the efforts introduced by standardization bodies such as IETF and IEEE.

IoT Communication Protocols & Their Security

Part of the standardization process of the IoT communications technologies is to identify protocols built to support internet communications with sensing devices, and to address the security requirements that must be met by such protocols. Constrained sensing platforms render most of the communications and security protocols deployed on the internet ill-suited for the IoT. Consequently, working groups laid new standards for IoT protocols that will play a fundamental role in enabling future IoT applications. Simultaneously, these solutions are being designed to maintain the interoperability with existing internet platforms.

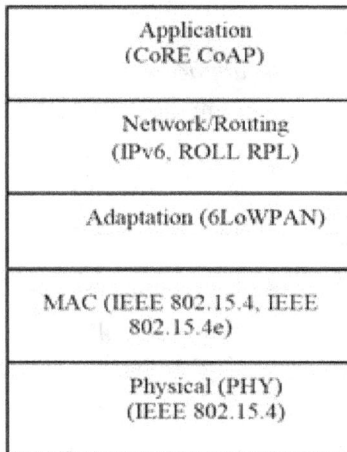

Application (CoRE CoAP)
Network/Routing (IPv6, ROLL RPL)
Adaptation (6LoWPAN)
MAC (IEEE 802.15.4, IEEE 802.15.4e)
Physical (PHY) (IEEE 802.15.4)

Figure (1) IoT Communication Protocol Stack

The protocol stack, shown in Figure 1, is designed with key features suitable to IoT platforms, the main characteristics of the protocols in this stack (bottom up) are:

1) IEEE 802.15.4 standard at the PHY and MAC layers supports low-energy communications and lays the ground for communication protocols at the higher layers.

2) Low-energy communication using IEEE 802.15.4 handles up to 102 bytes for the transmission of higher layers payload which is much less than the Maximum Transmission Unit (MTU) of 1280 bytes required by IPv6 packets. However, the 6LoWPAN adaptation layer overcomes this issue by translating IPv6 packets into IEEE 802.15.4 standard packets using packet fragmentation and reassembly.

3) Routing protocol for low-power and lossy networks (RPL) supports the routing over 6LoWPAN platforms. RPL is designed to be adaptive to a particular application requirements, rather than being a standard routing protocol.

4) The *constrained application protocol* (CoAP) [2] is proposed to support communications at the application layer. Currently, the IETF is designing CoAP for interoperation conformance with the representational state transfer architecture of the World Wide Web.

Security of IoT communications may be addressed in the context of the communication protocol itself. A communication protocol must provide assurances of various security requirements; such as integrity, authenticity, non-repudiation, and confidentiality, of the information flows.

In the following sub-sections, we identify how these requirements are dealt with by various security protocols at different layers of protocol stack shown in Figure 1. We analyze the security characteristics of these protocols. Moreover, we address the open issues each one may have and state the different proposals in the literature to overcome those issues.

Security in IoT PHY & MAC Layer

IEEE 802.15.4 standard is designed to support a trade-off between energy-efficient sensing platforms, transmission range, and data rate of communications. It supports data communications at 250 Kbit/s within a range of 10 meters. Despite IEEE 802.15.4 (2011) security is available at the MAC layer and it supports security mechanisms designed at higher layers of the protocol stack. This is motivated by the support of efficient symmetric cryptography in IEEE 802.15.4 hardware [3].

Security Modes supported by IEEE 802.15.4 MAC layer:

The security modes supported by IEEE 802.15.4 are listed in Table 1; they are distinguished by the guarantees they provide as well as the size of the integrity data employed. The level of frame secrecy could be determined by the mode being used, as indicated in Table 1. Moreover, the *security enabled bit field* of the *frame control field* identifies whether the frame is secured. When enabled, the *optional auxiliary security header* field is used which determines which security mode is selected, using the *security control field* inside. Likewise, the cryptographic key required for achieving the security process can be declared implicitly or explicitly. Figure 2 depicts the detailed hierarchical structure of IEEE 802.15.4 frame with all the security related fields.

Table (1): IEEE 802.15.4 Security Modes

Security Mode	Security Provided	
	Encryption	Authentication
No Sec	No	No
AES-CBC-MAC-32	No	Yes, using 32-bit MIC
AES-CBC-MAC-64	No	Yes, using 64-bit MIC
AES-CBC-MAC-128	No	Yes, using 128-bit MIC
AES-CTR	Yes	No
AES-CCM-32	Yes	Yes, using 32-bit MIC
AES-CCM-64	Yes	Yes, using 64-bit MIC
AES-CCM-128	Yes	Yes, using 128-bit MIC

Fundamental Security Requirements:

Here, we discuss the fundamental security requirements and how they are assured by the security fields' configuration of the IEEE 802.15.4 frame:

Confidentiality:

As mentioned earlier, security modes at the MAC layer are optional. A given communication node could only rely on the security protocols at the higher layers of the protocol stack.

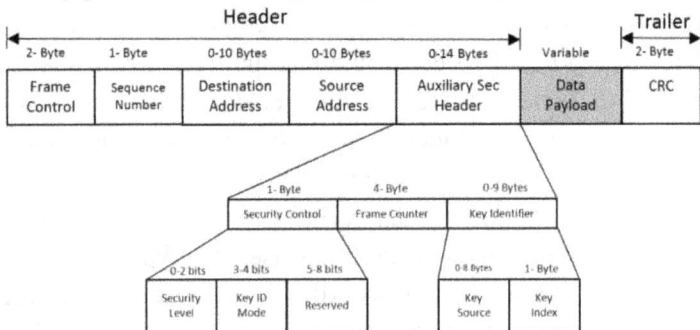

Figure (2) Security data and control fields in IEEE 802.15.4

On the other hand, it could merely rely on confidentiality provided at the MAC layer; where transmitted data is encrypted using the AES-CTR security mode with 128-bit keys.

Authentication and Integrity:

They are achieved using the AES in Cipher Block Chaining (CBC) mode, which produces a Message Integrity Code (MIC) or Message Authentication Code (MAC) attached to the transmitted data. Having different sizes of the MICs results in various security modes: AES-CBC-MAC-32, AES-CBC-MAC-64, and AES-CBC-MAC-128. Furthermore, in all these modes the payload is sent unencrypted. MIC code is created with information from the frame header as well as the payload data.

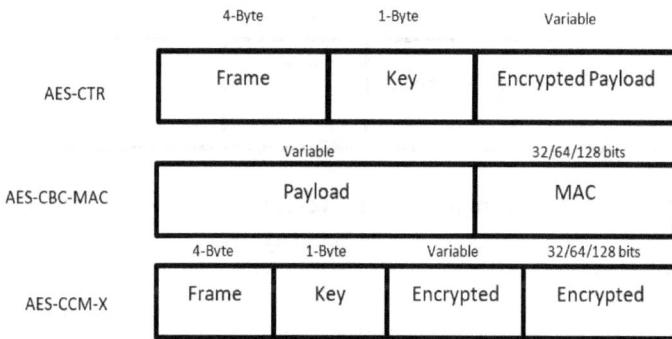

	4-Byte	1-Byte	Variable	
AES-CTR	Frame	Key	Encrypted Payload	

	Variable	32/64/128 bits
AES-CBC-MAC	Payload	MAC

	4-Byte	1-Byte	Variable	32/64/128 bits
AES-CCM-X	Frame	Key	Encrypted	Encrypted

Figure (3) Payload data formats with IEEE 802.15.4 security.

It is also possible to enable the Counter (CTR) and CBC modes jointly, using the combined Counter with CBC-MAC AES/CCM encryption mode; this would support confidentiality, message authenticity and integrity all together for MAC layer communications. The security modes are various and could be any of AES-CCM-32, AES-CCM-64, and AES-CCM-128, which, as mentioned earlier, differ on the MIC code size, following each message. Figure 3 shows a general case of AES-CCM modes which involves the transportation of all the security related fields after the encrypted payload.

Protection against Replay Attacks:

In any security mode, *frame counter* and *key control* subfields of the *auxiliary security header field* may be set to support semantic security and protection against message replay attacks. *Frame counter* subfield may be used to set the unique message identifier, and the *key counter* may be under the control of an application, which may increment it if the maximum value of the *frame counter* is up.

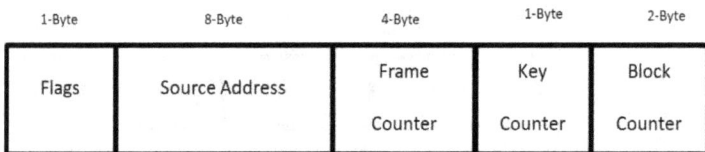

1-Byte	8-Byte	4-Byte	1-Byte	2-Byte
Flags	Source Address	Frame Counter	Key Counter	Block Counter

Figure (4) Format of the IV vector for AES-CRT & AES-CCM security

Moreover, the sender breaks the original packet into 16-byte blocks, called *initialization vectors* (IV), depicted in figure 4, and each of those blocks is identified by its *block counter*. To support semantic security and protection against message replay, each block is encrypted using a different IV. The Block Counter of the IV is not transmitted along with the message, instead it is inferred by the receiver. The IV is also used for encryption using various security modes of AES/CCM described earlier.

Access Control:

For access control, a sensing device may use the source and destination addresses of the frame to look up security related information required to process the security of the message. The IEEE 802.15.4 hardware stores *access control lists* (ACL) with up to 255 entries. Each of these entries contains the security-related information for a particular destination. Figure 5 shows the format of the IEEE 802.15.4 ACL.

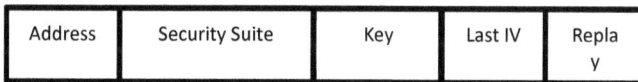

Address	Security Suite	Key	Last IV	Repla y

Figure (5) Format of an ACL entry in IEEE 802.15.4

An ACL entry stores a destination device address, a *security suite identifier field* and the security material needed to process security for communications with that device. This material includes the cryptographic key and, if encryption is enabled, the IV vector that must be preserved across different packet encryption invokes. When replay protection is in action, the *replay counter field* stores a watermark of the most recently received packet's identifier.

Security issues in PHY and MAC:

Despite the maturity of the IEEE 802.15.4 standard, various security limitations exist regarding how it implements the security services supported by the MAC layer of the communication protocol stack. Using IEEE 802.15.4 platforms for communications of the IoT raises some challenges and opportunities for further research work. Here, we discuss some existing proposals and opportunities that need future research and investigations:

- According to the IEEE 802.15.4 standard specifications, there is no key management mechanism specified. Moreover, this comes from the presumption that key management models are dependent on the threat model applicable to a particular application, as well as on the availability of the resources on the sensing device to support key management protocols.
- Mismanaging the IV values or the loss of ACL state after a node power interruption could result in using the same key in two or more ACL entries, this could be problematic as it is possible that the sender device will accidentally reuse the IV value. This situation is dangerous with stream ciphers in CRT mode, like AES/CCM, as it may permit an adversary to recover plaintexts from ciphertexts.
- Group keying and network shared keying may not be adequately supported by IEEE 802.15.4 tables, where ACL entries are stored. That comes from the fact that each entry may be associated with only a single destination address, which makes it difficult for group keying to be implemented. The support of group keying requires more than one ACL entry using the same key. On the other hand, network shared keying in not compatible with the protection against replay attacks. It may be supported using default ACL entry, transmitting nodes have to coordinate the usage of their replay counter space.
- Acknowledgment messages are not protected to assure integrity or confidentiality. An adversary may, therefore, forge acknowledgments, by learning the frame sequence number of the packet to perform DoS attack.

Those limitations need improvements in future works towards building new standards, or may also call for cross-layer interaction with other adopted security mechanisms at other layers of the protocol stack.

However, there exist some improvement proposals for IEEE 802.15.4 MAC security; like those that suggest designing key management protocols (KMP) at the higher layers, to support end-to-end security, thus overcoming the limitations of ACL management at the link layer in the context of supporting the group and network shared keying. KMP may also be designed to exploit the ACL storage available at IEEE 802.15.4, even without supporting MAC security. Likewise, since AES/CCM at the sensing devices hardware already supports efficient cryptographic basis, protocols at higher layers may also exploit this feature.

Security in IoT Network Layer

6LoWPAN is a key technology that supports internet communications on the IoT platforms. It has changed the perception of IPv6 as being impractical for constrained low energy WSN. The 6LoWPAN adapts the services required by the network layer with the services provided by the IEEE 802.15.4 MAC layer; thus, enables IPv6 end-to-end communications between constrained IoT sensing devices and other powerful internet entities, providing the required support for future IPv6-based distributed sensing applications on the IoT.

Security in 6LoWPAN:

According to the RFCs, there are no security mechanisms currently defined for the 6LoWPAN adaptation layer, but some documents include discussions on the security vulnerabilities, requirements, and approaches to consider for the usage of network layer security; as we discuss later. The current 6LoWPAN RFCs only discuss general security threats and requirements, despite RFC 4944 [4] clearly identifing the interest of adopting appropriate security mechanisms for the 6LoWPAN adaptation layer. Next, we present proposals in the literature for the protection of IoT network layer using 6LoWPAN:

Confidentiality, Integrity, Authentication, and Non-Repudiation:

In the current internet architecture, IPSec protocol would provide authentication and encryption, at the network layer, for a given communication session. Network layer security may provide a valuable usage in future IoT applications. So far, there are no specific security mechanisms adopted for the 6LoWPAN adaptation layer. The challenges are related to the resource constraints of the wireless sensing platforms.

On the other hand, a few research proposals exist for this purpose, targeting the design of a compressed security header for the 6LoWPAN adaptation layer [5]–[7]. Initially, the authors in [8] discussed that the use of compressed security headers at the adaptation layer is a considerable solution, as long as those headers are efficiently designed, and also the sensing devices support efficient hardware security optimizations.

Later, they proposed and evaluated the usage of AH, and ESP compressed security headers for 6LoWPAN in tunnel and transport modes [9], [10], considering predefined application security profiles and AES/CCM encryption at the hardware level. Also, in [11], the authors consider the design of compressed headers for 6LoWPAN. Another work in [12], presents an experimental evaluation of this proposal and its comparison with IEEE 802.15.4 link layer security.

However, those proposals would impose support from external internet entities, either by upgrading existing IPSec platforms to support 6LoWPAN compressed headers or by designing security gateways to support end-to-end network security between various security platforms.

Protection against Packet Fragmentation Attacks:

Due to the lack of authentication at the 6LoWPAN adaptation layer, packet fragmentation attacks could be problematic. Those attacks are executed against the 6LoWPAN fragmentation and reassembly mechanisms. However, authors in [13] discuss the consequences of such attacks. Their paper also proposes the addition of new fields to the 6LoWPAN fragmentation header to tackle such threats, employing hash chains to enable a legitimate sender to add an authentication token to each fragment.

Another solution suggests using mechanisms that support per-fragment sender authentication and purging of messages from the receiver's buffer, for transmitter devices considered suspicious, [14]. Unlike the former, in the latter, based on the observed sending behavior, the receiver decides which fragments to discard in case a buffer overload occurs. The receiver decision is based on per-packet scores, which capture the extent to which a packet is completed along with the continuity in the sending behavior.

Key Management:

Key management is an important security feature to discuss in the 6LoWPAN context. There are no specific solutions proposed for 6LoWPAN. However, RFC 6568 [15], discusses the possibility of adopting simplified versions of existing Internet key management solutions. For instance, in [16] minimal IKEv2 adapts current internet key management to constrained sensing platforms.

Another approach considers compressing the IKE headers and the payload information, [17]. In [18] the authors claim that public key management methods still need a more powerful sensing platforms. They also discuss the possibility of employing mathematical based key management solutions to support IoT applications.

Security for Routing in the IoT:

The current approach to handling routing in 6LoWPAN platforms is realized in RPL. RPL provides a framework adaptable to the requirements of particular applications, rather than providing a generic routing approach. Here, we discuss the security mechanisms designed to protect RPL communications.

The RPL specifications in [19] identify various secure-routing control messages, and three basic security modes as well. The format of a secure RPL control message is shown in Figure 6-A. To enable security, the high order bit of the *code field* is marked, to result in one of the various possible secure messages (DIS, DIO, DAO or DAOACK). The *security field* format is shown in Figure 6-B.

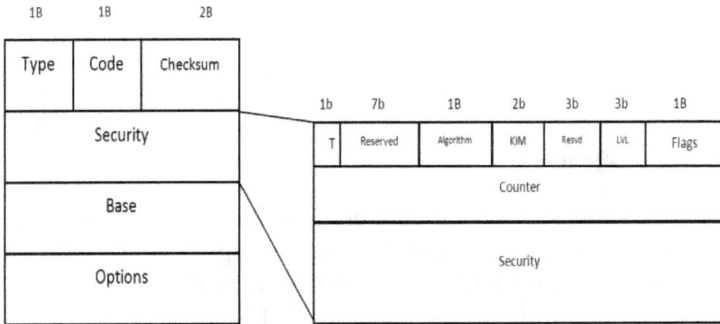

Figure (6-A) Secure RPL   Figure (6-B) Security Section Of
Control Message   Secure RPL Control Message

As shown in Figure 6-B, the Security field includes the necessary information required to define the security level and also the algorithms used to ensure a secure message. However, it does not include explicitly the security related data needed to process the security of the message; like, MIC Code or a signature. Different configurations of RPL would support various security features as explained in the following:

Confidentiality:

Confidentiality and delay protection could also be supported in RPL control messages. RPL control messages may be protected using either an integrated encryption & authentication suite, like AES/CCM, or schemes that handle encryption and authentication separately.

However, when an RPL message is encrypted, encryption is applied to the first byte after the security field and continues to the last part of the packet. In other words, the RPL message header up to the start of the *security field* is left unencrypted, as these fields are required to decrypt the packet at the receiving end.

Integrity and Data Authenticity:

According to RPL specifications in [19], integrity and data authenticity are defined by the use of AES/CCM with 128-bit keys for MAC generation, and of RSA with SHA-256 for digital signatures. The *LVL field* determines the level of security applied. Furthermore, RFC 6550 also defines various values to support confidentiality, integrity and data authenticity with MAC-32 and MAC-64 authentication codes, and also of 2048 bits and 3072 bits signatures using RSA.

Protection against Replay Attacks:

A sensing device may issue a challenge-response using a *consistency check (CC) control message*, to confirm another node's current counter value. For instance, when a received message has a certain initialized counter value and the receiver has an incoming counter currently maintained for the message originator, then the receiver initiates counter resynchronization by sending a CC message to the source.

Protection against packet replay attacks is enabled utilizing the *counter field*, where it is used to transport a timestamp T, see Figure 5-B. The next byte in the *security field* of the RPL control message defines the security suite used. On the other hand, the *flags field* is currently reserved.

Key Management:

The *key identifier mode* (KIM) subfield of the *security field*, Figure 5-b, indicates whether the security key is declared explicitly or implicitly. RFC 6550 [19] defines the various key management approaches that could be used: like group keys, keys per pair of sensing devices, and digital signatures.

Moreover, the key field is subdivided into the *key source* and *key index* subfields. The *key source* subfield specifies the logical identifier of a group key originator. On the other hand, the *key index* subfield allows unique identification of keys with the same originator.

RPL Security Modes:

RPL defines the following three security modes:

- Unsecured mode: This is the default mode of RPL, where there is no security applied to the routing control messages.
- Preinstalled mode: In this mode, a sensing device would have a preconfigured symmetric key so that it could associate with an existing RPL instance, as a host or a router.
- Authenticated mode: This mode is suitable for routers. A router may initially associate with a network with a preinstalled key and security mode, and then attain a different key from a key authority.

However, RFC 6550 [19] doesn't specify how asymmetric cryptography may be used to support authentication and key retrieval for a sensing device intending to operate as a router. More clarification on such mechanisms is required.

Other documents include some analysis on RPL security aspects; they introduce additional security mechanisms. For example, RFC letters [20]–[23] discuss routing requirements for the various application areas. They discuss the importance of protecting routing control messages with appropriate confidentiality, authentication, and integrity. RFC 6551 [24] identifies a set of link and node routing metrics suitable to the 6LoWPAN platforms and discusses the necessity of handling such metrics in a secure and trustful manner.

Security in IoT Application Layer

In IoT, application layer communications are supported by CoAP [2] protocol. CoAP protocol compresses, using a set of techniques, application layer data without compromising application interoperability. It is currently defined only for UDP communications over 6LoWPAN, although the adoption of transport layer approaches with characteristics closer to protocols such as the Transmission Control Protocol (TCP) is still underway with ongoing research [25].

Security in CoAP:

CoAP protocol defines bindings to Datagram Transport Layer Security (DTLS) [26] to secure CoAP data messages, along with a few mandatory minimal configurations suitable for constrained platforms. The integration of DTLS into CoAP communications is intended to support confidentiality, data authenticity, integrity, non-repudiation and protection against replay attacks.

However, it is apparent that DTLS security is maintained at the transport layer, rather than being only confined in the context of the application layer. Data communications on CoAP can be secured using DTLS. DTLS is originally TLS with some add-ons to combat the unreliability in UDP. CoAP payload space within DTLS is shown in Figure 6, once the initial handshake is accomplished, DTLS adds a 13 bytes overhead. 6LoWPAN header compression requires 10 bytes for UDP/IPv6 header, and CoAP fixed header requires 4 bytes.

Like other security approaches in 6LoWPAN platforms, AES/CCM algorithm is employed to support fundamental security requirements in the current CoAP specifications [2]. Security against replay attacks may also be achieved in the context of DTLS, using a different IV value for each CoAP packet.

CoAP Security Modes:

CoAP currently defines four security modes. Those security modes vary on how key negotiation and authentication are performed, as follows:

- No-sec mode: it provides no security, and CoAP messages are transmitted as they are.
- Pre-shared key mode: for devices that possess preconfigured keys and are required to communicate securely with other devices. This mode may also be suitable for devices that can't support public key cryptography.
- Raw public key mode: This security mode is defined as mandatory to implement in CoAP. It is suitable for platforms that require public key based authentication, but can't afford the process of calculating a key by negotiation with public key infrastructures. A given device must be preconfigured with a key pair. The device can derive its identity from its preconfigured public key. It also has a list

of identities and public keys of the nodes it can communicate with.

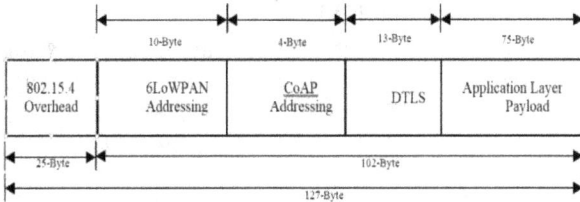

802.15.4 Overhead	6LoWPAN Addressing	CoAP Addressing	DTLS	Application Layer Payload

Figure (7) Payload Space with DTLS on 6LoWPAN Environments

- Certificates: Unlike raw public key mode, this mode supports authentication based on public keys for devices that are capable of participating in a certification chain with public key infrastructures. The device has a key pair with an X.509 certificate that binds it to its authority name and is signed by some common trusted root. Also, the device has a list of root trust-anchors with whom it validates its certificate.

An important feature of CoAP security using DTLS is the adoption of *Elliptic Curve Cryptography (ECC)* to support the security modes *Raw Public Key*, and *Certificates* [27]. ECC supports device authentication by using the *Elliptic Curve Digital Signature Algorithm (ECDSA)*, and also *key agreement* using the *ECC Diffie-Hellman counterpart*, the *Elliptic Curve Diffie-Hellman Algorithm with Ephemeral keys (ECDHE)*. Furthermore, CoAP specifications impose a mandatory implementation of cipher suite for each security mode, based on the usage of AES/CCM and ECC cryptographic operations, as follows:

- Applications that support pre-shared key mode security mode should at least support

(TLS_PSK_WITH_AES_128_CCM_8) suite. This suite provides authentication using pre-shared symmetric keys and 8-byte IV values, it also encrypts and produces 8 bytes integrity codes [28].

- Applications that support raw public key mode should support the *(TLS_ECDHE_ECDSA_WITH_AES_128_CCM_8)* suite using *ECDSA* capable public keys. This security mode also uses *SHA-256* to compute hashes [29], [30].

- Applications that support certificates security mode should support *(TLS_ECDHE_ECDSA_WITH_AES_128_CCM_8)* suite. The *Subject Public Key Info field* in *X.509* certificate declares how the corresponding public key must be used for ECC computations. The certificate also contains a signature created by *ECDSA* and *SHA-256*. For the applications that use devices with a shared key plus a certificate, they must also support *(TLS_ECDHE_PSK_WITH_AES_128_CBC_SHA).*

In addition to the cipher suites discussed, we may expect further suites to be implemented in future CoAP versions, to enhance the adaptation of the various security modes to various applications and sensing platforms. CoAP also doesn't define any mechanism for key management; it only assumes that initial keys are available from the DTLS authentication handshake.

Research Challenges and Proposals for Application Layer Security:

As discussed earlier, CoAP security at the application layer is supported by DTLS. It can be observed that DTLS has some limitations; this has motivated many ongoing types of research (e.g. CoRE, and DICE working groups) to work on new solutions to strengthen security at the application layer of IoT communication platforms.

Limitations of CoAP Security

DTLS could have a major impact on constrained wireless sensing platforms, and this comes from the cost of supporting the initial handshake as well as the processing of security for each exchanged CoAP message. This issue motivates many researchers to present alternative approaches to protecting IoT communications at the application layer using CoAP.

However, it is important to evaluate this impact on the sensing platforms with different characteristics because the DTLS handshake can exhaust the resources of those constrained devices, particularly considering the adoption of ECC public key cryptography for authentication and key agreements.

There is a big interest for investigating and optimizing DTLS in IoT environments, and also for conducting interoperability testing of DTLS implementations using 6LoWPAN and CoAP [31], [32].

Also, various features of DTLS protocol have been identified as posing challenges to its adoption in constrained sensing environments:

- The DTLS handshake could be difficult to support because large CoAP messages need fragmentation at the 6LoWPAN adaptation layer and the cost of the computation of the message at the end of the handshake is high [33], [34]. Fragmentation may also include retransmission and reordering of handshake messages which may result in more complexity.
- ECC employment on constrained sensing platforms is still not consensual. So, ECC support on 6LoWPAN platforms requires more investigation.
- For the validity of devices X.509 certificates, online verification mechanisms need further investigations, particularly for the CoAP Certificates mode.
- DTLS is not well suited to the usage of CoAP proxies in forward or reverse modes.

- DTLS is not able to support multicast communications, which will be required in many IoT situations [33], [34]. Also, secure CoAP multicast transmissions will require a suitable group key establishment mechanism.

Research Proposals:

Here, we present the research proposals promoting the effectiveness of DTLS to protect CoAP communications, and also, other approaches to support security for IoT application layer communications:

- Key Management: as discussed earlier, DTLS does not have a group key management mechanism. However, [36] proposes the adaptation of the DTLS record layer to enable more than one sender, in a multicast group, to send secure CoAP messages with a common group key, while maintaining message confidentiality, integrity, and replay protection. Authors in [35] assume that the required group keying material is available in the context of a given group security association, particularly the appropriate client and server read and write MAC keys, encryption keys, and IV values.
- DTLS Modification: some other DTLS features may not be suitable for IoT platforms, authors in [36] discuss some of those issues; for instance, the inadequateness of the timers used for message retransmission, as they may require buffering on the receiver side. Also, the efficiency of the code needed to support DTLS in IoT platforms. The authors also discuss the employment of stateless compression of the DTLS headers to reduce the overhead of DTLS records and handshake messages. Authors in [37] follow up on the work in [36], and propose the compression of the DTLS headers using LOWPAN_IPHC 6LoWPAN header compression. Another approach, in [38], suggests that CoAP should handle the heavy DTLS handshake operations. The authors of this proposal suggest a RESTful DTLS handshake to handle the problem of message

fragmentation at the 6LoWPAN adaptation layer. Their mechanism allows efficient transmission of handshake messages in the payload of CoAP messages, using blockwise transfers when required for larger messages.

- Offloading Costly DTLS Operations: Some proposals suggest the employment of gateways to support related security mechanisms in the context of DTLS communications. According to [33], [34], one issue to be overcome for CoAP security is the unavailability of a mechanism for mapping between DTLS and TLS. On the same issue, in [39], authors propose using a security gateway. Those gateways may also be adopted for mapping between CoAP and HTTP. Another proposed approach is to offload heavy DTLS operations to those gateways, as they are more powerful devices. However, few proposals consider this approach, focusing particularly on the entrustment of operations performed in the context of the DTLS handshake.

Authors of [40] propose a mechanism to support sleeping devices, using a mirroring mechanism to serve data on behalf of sleeping smart objects. The authors in [41] propose an end-to-end architecture supporting mutual authentication with DTLS, using specialized trusted platform modules supporting RSA on sensing devices, instead of ECC required by CoAP. Authors in [42] utilize a security gateway to transparently intercept DTLS handshakes between the CoAP client and server, which would offload ECC computations to the gateway. After the initial handshake, the gateway will be in possession of the keying material; it may be used to decrypt sessions between the CoAP parties, thus supporting additional security mechanisms involving traffic analysis.

- Public Keys Support & Digital Certificates: The processing of certificates using IoT platforms is an issue that requires proper performance evaluation. Authors in [43] discuss approaches to address this concern by considering the employment of a security intermediary. Their proposals are certificate pre-validation and session resumption. In Certificate pre-validation, a security gateway would be used to validate the certificates, in the context of the handshake, before forwarding the messages to a destination. On the other hand, session resumption lets sensing devices keep minimal session state after session teardown, which they may use later to resume secure communications without starting over new DTLS handshake. For very constrained sensing devices, this proposal addresses the full handing over of the DTLS handshake to a proxy.

- Object Security With CoAP: Alternative approaches, to secure CoAP communications, are being considered, in particular, approaches that employ object security rather than transport layer security. This solution may be realized by building security in CoAP protocol itself. Authors in [44] discuss the employment of three new CoAP options to provide security. The first option identifies how security is being applied to a given CoAP message and also the entity responsible for the processing of the message security. The second option is to enable the transportation of the necessary security data required for the authentication and authorization of a CoAP client. The third option is to enable the transportation of security data required to process cryptography for a CoAP message.

However, despite the research proposals presented, various research challenges in CoAP security require addressing; like, the lack of appropriate key management protocols for supporting CoAP multicast communications. Group key management mechanisms may be designed, either to be handled outside the CoAP protocol, or integrated with the DTLS handshake.

Also, for the usage of DTLS header compression, an appropriate support will also be required. On the other hand, mechanisms for mapping between compressed DTLS and DTLS may be designed, supported by security gateways. Security gateways may also support intrusion detection and attack tolerance mechanisms.

For intrusion detection for the sensing platforms, some existing works [45]–[47] may provide useful guidance for developing suitable mechanisms for 6LoWPAN based IoT communications.

Future work may also discuss the support of public keys and certificates in the context of CoAP security. Certificates online validation may be realized by investigating the applicability of existing internet approaches such as *online certificate status protocol* [48] or OCSP stapling through the TLS *certificate status request extension* defined in RFC 6066 [49], considering that such mechanisms could be integrated to support constrained 6LoWPAN platforms.

Another important issue is the computational burden of ECC on current sensing devices. Optimizations at the hardware of sensing platforms may be designed to support ECC costly computations, similarly to the support of AES/CCM by IEEE 802.15.4 platforms.

REFERENCES

[1] J. Sathish Kumar, and Dhiren R. Patel. A Survey on Internet of Things: Security and Privacy Issues. International Journal of Computer Applications (0975 – 8887) Volume 90 – No 11, March 2014.

[2] C. Bormann, A. Castellani, and Z. Shelby. CoAP: An application protocol for billions of tiny Internet nodes. IEEE Internet Comput., vol. 1, no. 2, pp. 62–67, Mar./Apr. 2012.

[3] IEEE Standard for Local and Metropolitan Area Networks-Part 15.4:Low-Rate Wireless Personal Area Networks(LR-WPANs). IEEE Std. 802.15.4-2011 (Revision of IEEE Std. 802.15.4-2006), (2011) 1-314, 2011.

[4] G. Montenegro, N. Kushalnagar, J. Hui, and D. Culler. Transmission of IPv6 Packets over IEEE 802.15.4 Networks. RFC 4944, 2007.

[5] S. Kent and K. Seo. Security Architecture for the Internet Protocol. RFC 4301, 2005.

[6] S. Kent and R. Atkinson. IP Authentication Header. RFC 2402, 1998.

[7] S. Kent and R. Atkinson. Encapsulating Security Protocol. RFC 2406, 1998.

[8] J. Granjal, J. Silva, E. Monteiro, J. Sa Silva, and F. Boavida. Why is IPSec a viable option for Wireless Sensor Networks. In Proc. 5th IEEE Int. Conf. MASS, 2008, pp. 802–807.

[9] J. Granjal, E. Monteiro, and J. Silva. Enabling network-layer security on IPv6 wireless sensor networks. in Proc. GLOBECOM, 2010, pp. 6–10.

[10] J. Granjal, E. Monteiro, and J. Silva. Network-layer security for the Internet of Things using TinyOS and BLIP. Int. J. Commun. Syst., vol. 27, no. 10, pp. 1938–1963, Oct. 2012.

[11] S. Raza, S. Duquennoy, and T. Voigt. Securing communication in 6LoWPAN with compressed IPsec. In Proc. Int. Conf. DCOSS Workshops, 2011, pp. 1–8.

[12] S. Raza, S. Duquennoy, J. Hoglund, U. Roedig, and T. Voigt. Secure communication for the Internet of Things-A comparison of link-layer security and IPsec for 6LoWPAN. Security Commun. Netw., vol. 7, no. 12, pp. 2654–2668, Dec. 2014.

[13] H. Kim. Protection against packet fragmentation attacks at 6lowpan adaptation layer. In Proc. ICHIT, 2008, pp. 796–801.

[14] R. Hummen et al. 6LoWPAN fragmentation attacks and mitigation mechanisms. In Proc. 6th ACM Conf. WiSec, 2013, pp. 55–66.

[15] E. Kim, D. Kaspar, and J. Vasseur. Design and Application Spaces for IPv6 over Low-Power Wireless Personal Area Networks (6LoWPANs). RFC 6568, 2012.

[16] H. René, H. Wirtz, J. H. Ziegeldorf, J. Hiller, and W. Wehrle. Tailoring end-to-end IP security protocols to the Internet of things. In Proc. 21st IEEE ICNP, 2013, pp. 1–10.

[17] R. Shahid, T. Voigt, and V. Jutvik. Lightweight IKEv2: A key management solution for both the compressed IPsec and the IEEE 802.15. 4 security. In Proc. IETF Workshop Smart Object Security, 2012, pp. 1–2.

[18] R. Rodrigo, C. Alcaraz, J. Lopez, and N. Sklavos. Key management systems for sensor networks in the context of the

Internet of things. Comput. Elect. Eng., vol. 37, no. 2, pp. 147–159, Mar. 2011.

[19] P. Thubert et al. RPL: IPv6 Routing Protocol for Low-Power and Lossy Networks. RFC 6550, 2012.

[20] M. Dohler, T.Watteyne, T.Winter, and D. Barthel. Routing Requirements for Urban Low-Power and Lossy Networks. RFC 5548, 2009.

[21] K. Pister, P. Thubert, S. Dwars, and T. Phinney. Industrial Routing Requirements in Low-Power and Lossy Networks. RFC 5673, 2009.

[22] A. Brandt, J. Buron, and G. Porcu. Home Automation Routing Requirements in Low-Power and Lossy Networks. RFC 5826, 2010.

[23] J. Martocci, P. De Mil, N. Riou, and W. Vermeylen. Building Automation Routing Requirements in Low-Power and Lossy Networks. RFC 5867, 2010.

[24] J. Vasseur, M. Kim, K. Pister, N. Dejean, and D. Barthel. Routing Metrics Used for Path Calculation in Low Power and Lossy Networks. RFC 6551, 2012.

[25] T. Zheng, A. Ayadi, and X. Jiang. TCP over 6LoWPAN for industrial applications: An experimental study. In Proc. IEEE 4th IFIP Int. Conf. NTMS, 2011, pp. 1–4.

[26] E. Rescorla and N. Modadugu. DTLS: Datagram Transport Layer Security. RFC 4347, 2006.

[27]SECG-Elliptic Curve Cryptography-SEC 1. http://www.secg.org.

[28] D. McGrew and D. Bailey. AES-CCM Cipher Suites for Transport Layer Security (TLS). RFC 6655, 2012.

[29] T. Dierks and E. Rescorla. The Transport Layer Security (TLS) Protocol Version 1.1. RFC4346, 2006.

[30] S. Blake-Wilson, N. Bolyard, V. Gupta, C. Hawk, and B. Moeller. Elliptic Curve Cryptography (ECC) Cipher Suites for Transport Layer Security (TLS). RFC 4492, 2006.

[31] IoT CoAP Plugtests. 28-30 Nov 2012, http://www.etsi.org/plugtests/coap2/ Home.htm

[32] 6LoWPAN Plugtests. 27–28 July 2013, http://www.etsi.org/news-events/events/663-2013-6lowpanplugtests

[33] O. Garcia-Morchon, S. Kumar, R. Hummen, and M. Brachmann. Security Considerations in the IP-Based Internet of Things. draft-garcia-coresecurity- 06, 2013.

[34] M. Brachmann, O. G. Morchon, S. Keoh, and S. Kumar. Security considerations around end-to-end security in the IP-based Internet of things. In Proc. Workshop Smart Object Security Conjunction IETF83, 2012, pp. 1–3.

[35] S. Keoh, S. Kumar, O. Garcia-Morchon, and E. Dijk. DTLS-Based Multicast Security for Low-Power and Lossy Networks (LLNs). draft-keohdice-multicast-security-08, 2014.

[36] K. Hartke. Practical Issues withDatagram Transport Layer Security in Constrained Environments. draft-hartke-dice-practical-issues-01, 2014.

[37] R. Shahid, T. Daniele, and T. Voigt. 6LoWPAN compressed DTLS for COAP. In Proc. 8th IEEE Int. Conf. DCOSS, 2012, pp. 287–289.

[38] S. Keoh, S. Kumar, and Z. Shelby. Profiling of DTLS for CoAP-Based IoT Applications. draft-keoh-dice-dtls-profile-iot-00, 2013.

[39] M. Brachmann, S. Keoh, O. G. Morchon, and S. S. Kumar. End-to-end transport security in the IP-based Internet of things. In Proc. 21st Int. Conf. Comput. Commun. Netw., 2012, pp. 1-5.

[40] M. Sethi, A. Jari, and K. Ari, End-to-end security for sleepy smart object networks. In Proc. 37th IEEE Local Comput. Netw. Workshops, 2012, pp. 964–962.

[41] T. Kothmayr, C. Schmitt, W. Hu, M. Brunig, and G. Carle. A DTLS based end-to-end security architecture for the Internet of Things with two way authentication. In Proc. 37th IEEE Conf. LCN Workshops, 2012, pp. 956–963.

[42] J. Granjal, E. Monteiro, and J., Sá Silva. End-to-end transport-layer security for Internet-integrated sensing applications with mutual and delegated ECC public-key authentication. In Proc. IFIP Netw., 2013, pp. 1–9.

[43] R. Hummen, J. Ziegeldorf, H. Shafagh, S. Raza, and K.Wehrle. Towards viablecertificate-based authentication for the Internet ofthings In Proc. 2nd ACM Workshop Hot Topics Wireless Netw. Security Privacy, 2013, pp. 37–42.

[44] J. Granjal, E. Monteiro, and J. Sá Silva. Application-layer security for the WoT: Extending CoAP to support end-to-end message security for Internet-integrated sensing applications. In Wired/Wireless Internet Communication. Berlin, Germany: Springer-Verlag, 2013, pp. 140–153.

[45] I. Butun, S. D. Morgera, and R. Sankar. A survey of intrusion detection systems in wireless sensor networks. IEEE Commun. Surveys Tuts., vol. 16, no. 1, pp. 266–282, 2014.

[46] M. Young and E. Boutaba. Overcoming adversaries in sensor networks: A survey of theoretical models and algorithmic approaches for tolerating malicious interference. IEEE Commun. Surveys Tuts., vol. 13, no. 4, pp. 617–641, 2011.

[47] A. Abduvaliyev, A. Pathan, Z. Jianying, R. Roman, and W. C. Wong. On the vital areas of intrusion detection systems in wireless sensor networks. IEEE Commun. Surveys Tuts., vol. 15, no. 3, pp. 1223–1237, 2013.

[48] M. Myers, R. Ankney, A. Malpani, S. Galperin, and C. Adams. X. 509 Internet Public Key Infrastructure Online Certificate Status Protocol-OCSP. RFC 2560, 1999.

[49] D. Eastlake. Transport Layer Security (TLS) Extensions: Extension Definitions. RFC 6066, 2011.

CHAPTER SIX

THREAT TAXONOMY FOR CLOUD OF THINGS

Md Sadek Ferdous, Raid Khalid Hussein, Madini O. Alassafi, Abdulrahman Alharthi, Robert J. Walters and Gary Wills

Electronic and Software Systems Group, Electronics and Computer Science, University of Southampton, SO17 1BJ, UK

ABSTRACT

In the last few years, the Cloud Computing paradigm has experienced considerable growth, making it the de-facto technology fuelling almost all major online services. At the same time, the concept of Internet of Things has started to gain mainstream traction with the promise to usher in a new era of pervasive sensing using a network of numerous inter-connected IoT devices. Such IoT devices can generate an enourmous amount of data which becomes increasingly difficult to process using the limited computational and storage capabilities of these devices. To tackle this problem, a notion of a novel technology called Cloud of Things is emerging.. However, to harness the full potential of this new paradigm, different security and privacy issues need to be properly analysed. The first step for carrying out such an analysis is to define a well-constructed threat model. In this chapter, we present a comprehensive threat model which is then utilised to create a first-ever threat taxonomy for Cloud of Things. This taxonomy outlines different security and privacy threats faced by this nascent technology and can be used as the basis for further research on security and privacy in Cloud of Things.

Keywords: *Cloud Computing, Internet of Things (IoT), Cloud of Things (CoT), Threat model, Threat taxonomy.*

INTRODUCTION

The advancement of Information and communication technologies (ICT) and the growing demand for ubiquitous computing has prompted the emergence of new technological paradigms such as Cloud Computing and Internet of Things (IoT). Due to the limited computation capabilities and storage in IoT devices, Cloud Computing is considered to be the desirable means for hosting and developing large-scale IoT service platforms. Integration of Cloud Computing and IoT has led to a novel technological trend called *"Cloud of Things (CoT)"*. CoT is *"a concept that provides smart things' functions as a service and allows them to be used by multiple applications"* 0. CoT can be leveraged to provide a processing, analysis, storage and decision making platform which utilises data sensed by IoT devices in a distributed architecture. In CoT, heterogeneous resources can be accumulated according to a unified processing technique, thus enabling creation of services which provide monitoring, analysis and visualisation of data sensed by multitude of IoT devices 0.

Since the concept of CoT is an integration of Cloud Computing and IoT technologies, several security and privacy challenges arise which are inherited from the architecture of the underlying technologies involved in both Cloud Computing and IoT. These security and privacy threats may hamper the growth of CoT 0. Therefore, their impact needs to be critically analysed. There exists a plethora of research papers in the existing literature that analyse different security and privacy issues with respect to Cloud Computing and IoT separately but none provides a comprehensive analysis of security and privacy threats in the setting of CoT. The concepts of security and privacy are probably related but they are different and need to be analysed separately in order to categorise the common threats associated with both paradigms.

This chapter provides an insight into CoT paradigm and its underlying concepts are critically reviewed in order to establish a comprehensive definition of CoT. In addition, a threat model is presented which outlines the assumed capabilities of an adversary (attacker) within a CoT system. Based on these properties of the attacker, different security and privacy threats have been identified and then combined to formulate a taxonomy of threats for CoT. Finally, a review of the identified security and privacy threats within the taxonomy is presented. In summary, the contributions in this chapter are as follows:

- A comprehensive definition for CoT.

- A threat model formulated for CoT by outlining an associated adversary model with different assets.

- A taxonomy of threats based on the threat model.

- Brief clarification of each identified threat.

The chapter is organised as follows. In Section 0 the concept of

CoT is investigated and a rigorous definition is presented. In Section 0, a threat model for CoT is outlined. A taxonomy of different security and privacy threats is presented in Section 0 along with the description of the identified threats and their corresponding attacks. The chapter concludes in Section 0.

CLOUD OF THINGS: CONCEPT AND STRUCTURE

In this section an overview of the notion of CoT is given and the origin of its underlying terms 'Cloud' and 'Things' is elaborated. Finally, a comprehensive concrete definition for CoT is presented based on a review of the visions, characteristics and architectures of Cloud Computing and Internet of Things (IoT).

A. Cloud Computing

Cloud Computing has evolved from technologies such as virtualisation, grid computing, distributed computing, web 2.0 technologies, Service Oriented Architecture and utility computing. There are many definitions for Cloud Computing in the literature. The three most cited definitions are presented below.

- *Buyya et al.* defined Cloud Computing as *"A type of parallel and distributed system consisting of a collection of interconnected and virtualized computers that are dynamically provisioned and presented as one or more unified computing resources based on service-level agreements established through negotiation between the service provider and consumers"* 0.

- *Vaquero et. al.* define Cloud Computing as: *"a broad array of web-based services aimed at allowing users to obtain a wide range of functional capabilities on a 'pay-as-you-go' basis that previously required tremendous hardware/software investments and professional skills to acquire"* [5].

- National Institute of Standards and Technology (NIST) defines Cloud Computing as *"a model for enabling convenient, on-demand network access to a shared pool of configurable computing resources (e.g., networks, servers, storage, applications and services) that can be rapidly provisioned and released with minimal management effort or service provider interaction"* 0.

Of these definitions, that by NIST is the most comprehensive as it encompasses the unique characteristics, service models and deployment models of the cloud.

A-1.   Cloud Characteristics

According to the NIST definition, Cloud Computing has five characteristics as follows:

- **On-demand-self-service:** In Cloud Computing, users can automatically utilise computing resources such as servers, software, and storage as desired without any human interaction with a cloud service provider [1].

- **Resource pooling:** The cloud provider's pool of computing resources is grouped together to serve multiple tenants/clients in such a way that different physical and virtual resources are automatically allocated and relocated according to the user's demands [2].

- **Broad network access:** Resources in Cloud Computing are reachable over the Internet by standard techniques and used by heterogeneous thin or thick consumers' platforms [3].

- **Rapid Elasticity:** Computing resources can be promptly and elastically scaled out and in depending on the demand for resources [4]

- **Measured Service:** Resources can be monitored, controlled, provisioned and charged according to a service level agreement which will ensure transparency for the cloud clients and the service providers [2].

A-2.   Cloud Service Models

Cloud Computing has three service models or three architectural layers as identified by NIST as in *Table 1*.

A-3.   Cloud Deployment Models

There are four deployment models for cloud services, with derivative variations that address specific requirements. The four models are listed below:

- **Public Cloud:** In this model, a single organization generally owns the infrastructure. The infrastructure is made available for public or other organizations and is leveraged to provide different services. This is currently the most widely used model globally [8]

- **Private Cloud:** In this model, the infrastructure is utilized by a single organization and hence, it is not made available to anyone outside the organization. The infrastructure can either be managed by the organization or another organization may manage it on behalf of the first organization [9]

- **Community Cloud:** In the model, the infrastructure is shared among multiple organizations which may share a set of common goals and requirements among themselves. The infrastructure is managed by different members of the community using a pre-determined level of agreement [10]

- **Hybrid Cloud:** In this model, the cloud infrastructure is a combination of two or more other cloud models where particular application scenarios prohibit the usage of a certain cloud model [8].

Table 1: Cloud Service Models

Services	Description
Software as a service (SaaS)	This is the highest layer and comprises a complete application layer offered as a service, on demand, via multi-tenancy. For example, Salesforce, Facebook, LinkedIn, Intuit, Google Apps and Microsoft Office Live offer basic business services such as e-mail and messaging using the SAAS model [5].
Platform as a Service (PaaS)	Consumers using PaaS can develop and/or deploy applications by using provider's services and tools. PaaS providers provide tools for every phase of software development and testing which can be utilised to deploy any service quickly. Examples include Google App Engine and Microsoft Azure [6].
Infrastructure as a Service (IaaS)	Offers a means of delivering basic storage and compute capabilities as standardized services over the network. Amazon (AWS) and Rackspace are IaaS providers which provide servers, storage and other computing resources [7].

B.   Internet of Things

The term Internet of Things (IoT) was introduced for the first time by the British entrepreneur Kevin Ashton in 1999 during his work at Auto-ID centres where it was defined as "*a universal network of devices connected via radio frequency identification (RFID)*" 0. The IoT concept seems difficult to understand because of the numerous visions, ideas and the diverse socio-economic and technical applications of IoT deployment.

However, researchers have highlighted the main concepts in the IoT model and clarified the vision of IoT in order to simplify the overall understanding of the IoT paradigm. European technology platform on smart systems integration (EPOSS) defined IoT as, *"world-wide network of interconnected objects uniquely addressable, based on standard communication protocols"* 0. In addition, Dunkels and Vasseur defined IoT as *"a world where things can automatically communicate to computers and each other providing services to the benefit of the human kind"* 0. Guillemin and Friess defined IoT as a *"dynamic global network infrastructure with self-configuring capabilities based on standard and interoperable communication protocols where physical and virtual 'things' have identities, physical attributes, and virtual personalities and use intelligent interfaces, and are seamlessly integrated into the information network"* 0. Atzori and Morabito state that in the IoT, the RFID (Radio Frequency Identification) is the corner stone of the technologies driving the vision 0. However, they indicate that a wide range of objects and technologies consisting of different types of networks, devices and electronic services are recognised as *"the atomic components that will link the real world with the digital world"* 0.

However, to ease the understanding of the definition of IoT, it can be broken down into two core concepts: Internet which refers to the universal network, and "Things" indicating a variety of hardware devices such as traffic monitoring devices, climate control devices, vehicles, appliances, etc. Some scholars view Things as a mixture of data, service, software and hardware 0. Therefore, IoT can be defined as a *"network of smart devices, vehicles and other items embedded with different systems, sensors, and network protocols, and application, which are capable of collecting and exchanging data".* This definition exhibits a sense about the architecture of IoT, which is often divided into three main layers in the literature: the perception layer which perceives (senses) the data from the environment, the network layer which is used to collect the data perceived by the previous layer and send it to the Internet, and the application layer which is used to create innovative applications, online services and real-life use-cases utilising the data collected and disseminated via the previous two layers 0, 0.

The novel paradigm of IoT has grown out the notion of computing beyond traditional computers. IoT objects exchange data with each other and other applications using wireless communication. It is envisioned that, in the furture, many objects around us will be equipped with the functionalities of IoT devices. For example, objects in domains such as e-health, learning, manufacturing automation, smart transportation and remote control for home appliances and leisure devices, may form a network of smart devices capable of collecting and sharing useful data 0. An example of IoT domains and potential applications is illustrated in Figure 1.

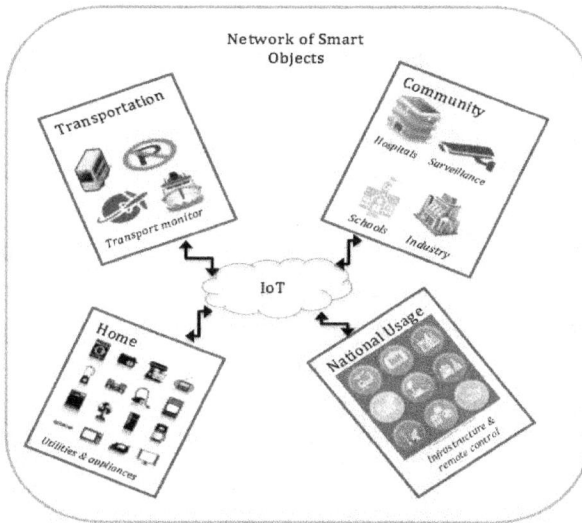

Figure 1: Internet of Things daily application areas

C. Cloud of Things

The number of the connected devices has already reached 9 billion and it is projected to reach 24 billion by 2020 0. Moreover, governments in Europe and Japan are already promoting "smart cities" in order to improve management of resources by using cutting edge Information and communication technologies (ICT). This utilisation of ICT will help cities to control the growing demands for increasing the quality of life, which lead to smarter life. The advancement in sensor and wireless technologies can enable monitoring and sensing the physical events using embedded computing and transmit the data wirelessly 0.

The growth in demand for data will lead to handling large

amount of data which will need scalable IT environments able to generate, transmit and process such data. Processing and analysing such data in innovative ways will generate novel knowledge. This high volume of data manipulation cannot be achieved using the limited processing and storing capabilities of existing IoT devices alone. Merging IoT with Cloud Computing provides a solid platform where a high performance pool of resources can be leveraged using a relatively low cost payment model (Pay-as-you-go) to host and process these huge volumes of data. The integration of these two paradigms leads to a novel paradigm called "Cloud of Things" (CoT). The vision of CoT is to introduce a set of heterogonous APIs of smart things as services to be visible and shared by other applications. The applications in the CoT platform can leverage (smart) things deployed by others without having to build their own infrastructure 0.

The CoT paradigm is commonly defined as *"an IT paradigm to connect physical objects to the cloud to meet the demand of huge real-time data"* 0, 0, 0, although Distefano *et al.* emphasise sensing as service concept when defining CoT 0. They define it as *"geographically distributed infrastructure with cloud agent elements that continuously discover and pool sensing resources of IoT devices to be used by cloud user's on-demand"*. In spite of the existing CoT definitions in the literature, there is no comprehensive concrete definition for CoT which can be derived by critically analysing the underlying IT concepts of both Cloud Computing and IoT.

Based on a critical review of the definitions of both IoT and Cloud Computing in previous sections, our synthesised comprehensive CoT definition is as follows:

"The Cloud of Things is a scalable IT paradigm for providing pay per use on demand network access to a self-configurable mutual pool of identified interconnected sensing devices embedded with different technologies (e.g., Wireless Sensor and Actuator Networks (WSAN), Applications, Near Field Communications (NFC), Radio Frequency Identifier (RFID)), which can be distributed globally and promptly provisioned in order to perceive data from the real world environments and link it with the digital world".

Based on this CoT definition, it can be understood that the concept of CoT differs from the concept of Cloud Computing. This is because simultaneous accesses to smart things (IoT devices) will need to be constrained in some cases to avoid resource conflicts so this conflict will not affect the workload of computing resources like memory, CPU, power, etc. For instance, a CoT-enabled projector in a room may need to restrict multiple access features simultaneously because the projector is able to show only one screen from a device (e.g., Smart Phone) at a time. To harness the full potential of this new paradigm, different security and privacy issues need to be properly analysed. The first step for carrying out such an analysis is to define a threat model which will outline the underlying threats faced by this nascent technology. In the subsequent sections, we present a threat model by creating a threat taxonomy for CoT along with a brief description of each of the identified threats.

THREAT MODEL IN CLOUD OF THINGS

The first step to design and develop a secure system is to formulate a threat model for the corresponding system. A well-defined threat model helps to identify threats to different assets of a system by utilising well-grounded assumptions of the capabilities of anyone interested to attack such a system. Then, in order to tackle such threats, different mitigation strategies are sketched out by formulating suitable security and privacy requirements 0. In summary, a threat modelling process consists of the following 0 0:

    i. illustrating the adversary model,
    ii. listing assets of the system,
    iii. identifying possible threats to those assets and
    iv. formulating mitigation strategies.

Since each system has different assets, the threat modelling process of one system within the domain of CoT will be considerably different from that of another system within the same domain. Even so, there are some common assets, hence some common threats, that will exist in any system within a particular domain, e.g., CoT. In this chapter, the focus will be mainly on the first three steps (analysing the adversary model, listing assets and identifying threats) of the threat modelling process and therefore, the discussion involving the fourth step (outlining mitigation strategies) is deliberately omitted.

The first two steps are described in the next section whereas a full section is devoted to the third step.

A. Adversary Model

An adversary model assumes certain properties and capabilities of an adversary (attacker) which the attacker may employ to attack a system. These assumptions are useful to underline what an attacker can and cannot achieve and are an essential ingridient for modelling threats. Indeed, the assumed capabilities of an attacker, in many ways, can influence the identification of assets which are exposed to the attacker. This, in turn, can be used to identify threats to these assets which can be exploited by the attacker.

There are different adversary models available with respect to the security of a system. However, two most widely-used adversary models are: Dolev-Yao (DY) model 0 and Honest-but-curious (HBC)/Semi-dishonest model 0, 0.

The DY model assumes that an adversary (also known as the DY adversary) has full control over the communication network in such a way that they can launch active and/or passive attacks. Active attacks include interrupting and altering messages, and replaying old messages as well as injecting forged messages into the communciation path. Passive attacks include eavesdropping and intercepting messages. These capabilities may enable the adversary mount attacks even without having any control of the respective system. The limitation that is imposed on the capability of such an attacker is that they cannot break secure cryptographic mechanisms. For example, the adversary cannot decrypt an encrypted message or falsify a digital signature without obtaining the necessary key(s), nor they can reverse cryptographic hash functions.

The HBC model assumes an adversary with limited capability (also known as the Semi-dishonest adversary) such that they cannot launch any active or passive attacks on the system. Instead, they may participate in the system as an honest user and follow its associated communication protocols as intended. However, the adversary will try to learn and infer as much knowledge as possible from just the messages intended for them. This may involve linking messages and analysing the contents of such messages. It is to be noted that the capabilities assumed for adversaries of both models mostly involve engaging in a communication protocol. However, we assume additional capabilities for both types of adversary which they can use to launch attacks in CoT systems. These additional capabilities are:

- Both types of adversary may have access to the physical environments where IoT devices are deployed. This may allow them to launch active or passive attacks within the respective physical environment.

- The adversaries may control substantial external computational power which resides out of the domain of the attacked system to launch attacks and/or infer knowledge.

- Adversaries may act as a malicious insider having expert knowledge on the deployment system which can be used to launch insder attacks.

B. Listing Assets

An asset is the abstract or physical resource in a system that needs to be protected from an adversary (attacker) 0. It is the resource for which a threat exists and represents the target of the adversary in the system. An asset can be a physical resource, for example, an item of hardware. Examples of such hardware are IoT devices which are used to sense and feed multitude of data from environments in which they are deployed. Another example is the cluster of server machines which is used to deploy cloud-enabled systems. Hardware may also include network components such as routers by which data is transmitted from one system to another and machines such as computers, smartphones and tablets from which users access different online services. An asset can also be an abastract resource including different software such as operating systems running the physical servers and hypervisors responsible for maintaing different virtual machines. Moreover, web servers, web services and data are also examples of software assets. Since the domain of CoT involve an amalgamation of IoT and Cloud technologies, both physical and abstract resources need to be considered. The motivation behind this step is to highlight those assets in the system which can be the target of an adversary so that associated threats for these assets can be identified. We identify the following physical assets:

- **IoT devices**: IoT devices are essential components of any CoT system. These are the physical devices that are deployed into the physical environment, to sense and collect data.

- **Cloud servers**: Cloud servers along with their associated storage capacities as deployed in different data-centre represent another crucial physical resouce. These devices are used to deploy different models of cloud services.

- **Client devices**: Client devices include computers, smartphones and tablets and are used to access services provides using the CoT systems.

We identify the following abstract resources:

- **Data generated by IoT devices**: Data generated by multitude of IoT devices deployed in different physical environments represent a crucial abstract asset. Such data can be extremely sensitive depending on the type of IoT devices and is one of the princicple assets that needs to be secured.

- **Identities of users and devices**: The (partial) identities of users and devices are crucial assets as these identities are used to identify and authenticate entities in a CoT System. Such identities can also be used to bind data with the corresponding entities and to ensure that data is generated from a valid entity.

- **Activities associated with an identity**: How the partial identity of an entity is leveraged for accessing different CoT services represents a valuable asset. This is because such associated activities offer an effective way for an adversary to track users across different application domains.

- **Web servers**: These servers are used to host CoT web-services. Such software needs protection to ensure only legitimate services are hosted.

- **CoT-enabled web services**: CoT-enabled web services may utilise extremely sensitive data generated from different sensors and hence, must be protected from unauthorised access. Such services should be exposed only to those entities which are properly identified, authenticated and authorised using their (partial) identities.

- **Cryptographic keys**: If crypographic mechanisms are used to ensure the security and privacy of different sensed data, the corresponding cryptographic keys represent a valuable assent. If these keys are compromised, the security and privacy of the system will break down.

- **Hypervisors & VM components**: Hypervisors are used to deploy and monitor different VMs whereas VMs are used to host web servers and web-services discussed above. Protecting hypervisors and VM components is crucial to ensure that these components are not compromised.

Threat Taxonomy

A threat is an activity or capability of an adversary against an asset of a system made or used with an intention to invade the security of the system or the privacy of a user in the system 0. Privacy is concerned with the ability to control who is able to access certain data (which may contain sensitive information regarding a person). On the other hand, security involves the mechanisms used to ensure the confidentiality, integrity and availability of the data at different points within a system 0. The motivation behind this step is to identify possible threats related to two categories: security and privacy.

Based on the adversary model and the assets, we have identified several security and privacy threats for any CoT system. These threats have been combined to formulate a novel taxonomy of threats for a CoT System. The taxonomy is illustrated in Figure 2, and is discussed below.

A. Security Threats

Threats that can compromise the security of the system belong to this category. These threats have been grouped together into the following sub-categories: communication threats, physical threats, data threats, service provisioning threats and other threats. Each of these sub-categories along with the identified threats is discussed below.

A-1. Communication threats

An attacker can abuse the communication channel between different entities within a CoT system to initiate threats belonging to this sub-category.

- **Availability:**

The threat to availability in a CoT system can be materialised using a Denial of Service (DoS) attack to prevent valid users accessing the respective services either by stopping the service or by exhausting and/or cutting down the communication channel and resources in such a way that a valid user cannot avail themselves of any service 0. In a DoS attack, a server in a CoT system (e.g., a server in a cloud cluster) can be flooded with service access requests. To cope with these requests, the operating system in the server may start to utilise additional computation power and resources. Since the server has limited processing capability, these requests exhaust all its resources leaving no resources to be utilised to perform tasks needed to service legitimate service requests. In this way, the attacker could block services by limiting the server's available capability 0.

In the traditional Cloud Computing setting, a DoS attack can affect the provider resources and the network as mentioned above. However, in a CoT system, the DoS attack can also affect the availability of the systems where data are produced by different IoT devices. The attacker can jam the wireless channel which in turn can affect the communication channel between the IoT devices and the cloud system and thus launching a DoS attack 0.

- **Eavesdropping:**

An eavesdropping attack allows an attacker to gain access to different communication channels and exploit such channels to extract the circulated data during interactions between different entities within a system in the targeted infrastructure [29, 30]. Such data may then used to breach the security of the system and to invade the privacy of the users. It is considered a major communication threat which affects the confidentiality of data during transmission.

- **Spoofing:**

A spoofing attack allows a malicious party to mimic another device or impersonate another user on the same communication medeium in order to spread malware or bypass access control mechanisms. There are several types of spoofing attacks. Two of the most common spoofing types are IP spoofing and ARP (Address Resolution Packet) spoofing. In the IP spoofing attack, one device mimics other devices by using their IP address. Whereas in the ARP spoofing attack, one device spoofs other devices using their MAC addresses 0.

- **Man-in-the-middle (MITM) attack:**

In the MITM attack, an attacker places herself in between a valid sender and receiver without their knowledge. Then, the attacker intercepts data packets in transmission between the sender and receiver and replaces valid data packets with fraudulent ones in such a way that the receiver is tricked into believing that these packets have been generated by a valid sender. This may enable the attacker to impersonate a legitimate sender 0. An example of a MITM attack scenario in a CoT system which deploys IoT sensors would be an attacker who interferes with temperature data from the sensor devices to induce a controller in the system to let the system overheat which could lead to physical and financial damage 0.

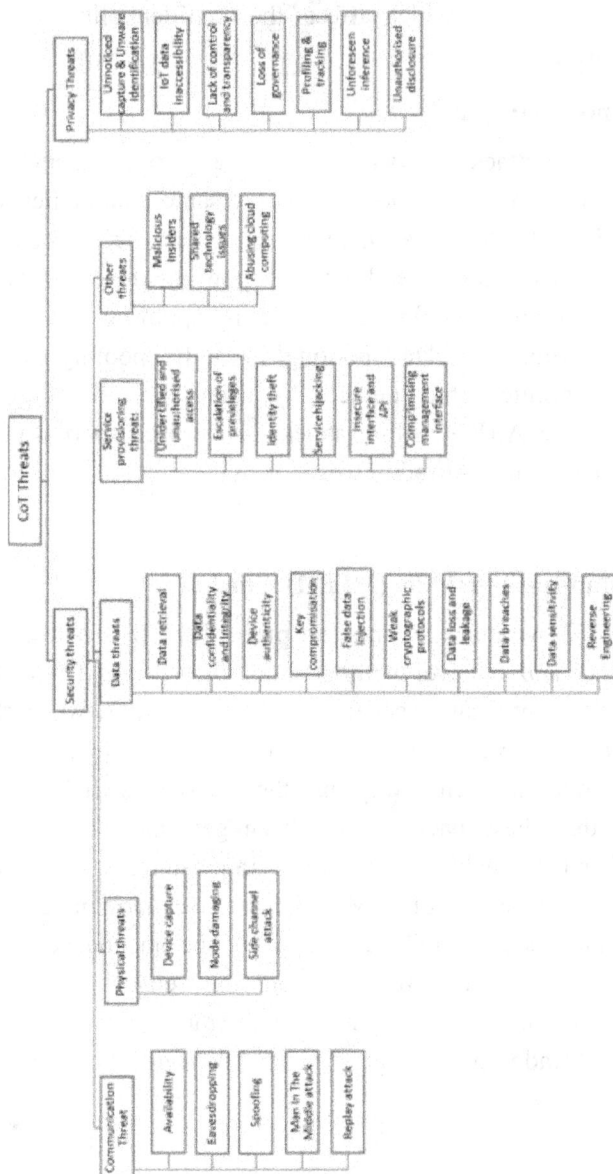

Figure 2: Taxonomy of threats for CoT

- **Replay attack:**

The replay or playback attack is similar to the MITM attack. It is a kind of communication attack in which a genuine message is intercepted and later maliciously replicated by the attacker. After receiving such a duplicate of a legitimate message, the receiver might be tricked in believing that the duplicate message is genuine and from a legitimate user 0. It differs from the MITM attacker as the attacker injects fraudulent messages in the MITM attack whereas in the playback or replay attack, the attacker simply resends a copy(s) of a legitimate message.

A-2. Physical threats

The threats that are materialised only if an attacker can compromise the physical IoT devices belong to this sub-category. The identified threats are described below.

- **Device capture:**

An attacker may gain access to the physical environment where different IoT devices are deployed. Gaining such access might enable the attacker to interfere with the operation of deployed devices, including extracting or changing information before it is sent into the system 0.

- **Node damaging:**

An attacker having physical access to the environment where IoT devices are deployed can physically damage a device, thus preventing it from sensing and transmitting data. If the attacker can damage a large number of devices, they may be able to deprive the system of input data, making the system unusable 0.

- **Side channel attack:**

An attacker may launch attacks such as timing attacks,

power analysis attacks, fault analysis attacks, electromagnetic attacks and environmental attacks which rely on physical characteristic of the devices and the environments in which they are deployed 0, 0. By examining these physical characteristics, certain level of inference can be made which ultimately can be exploited to compromise the security of the system. Such attacks rely on the way that logical functionalities may exhibit certain deterministic physical behaviour and characteristics depending on the input data and communication pattern. Using such mechanisms, the attacker may even be able to attack cryptographic mechanisms that have been deployed to protect the system 0.

A-3.   Data threats

Threats and their corresponding attacks which rely on data generated by a

CoT system belong to this sub-category. The identified threats and attacks are presented below.

- **Data retrieval from devices:**

When an attacker is able to gain physical access to any IoT device, the opportunity is likely to arise for them to retieve raw sensed data by tampering the device, reverse engineering or micro-probing 0. Data collected in this manner can be used to invade the security and privacy of users within the environment where the device has been deployed.

- **Data Integrity & Confidentiality:**

Once data has been collected by an IoT device, it needs to be transferred to a CoT system for storage and further analysis. This imposes the risk of data tampering while it is transmitted and stored 0. Therefore, the integrity and confidentiality of such data must be ensured while in transmission as well as during storage.

- **Device authenticity:**

An attacker may try to deploy unauthorised devices in the environment which might generate fraudulent data. To guarantee the trustworthiness of any CoT system, data should only be accepted from properly authenticated and authorised devices 0.

- **Key compromisation:**

An attacker can try to compromise the keys that are used in cryptographic mechanisms to guarantee the security of data. Getting hold of such keys enables the attacker to compromise the security of CoT systems and their respective users.

- **False data injection:**

An attacker may deploy an unauthenticated and unauthorised IoT device, or an authenticated and authorised IoT device which has been compromised in order to inject false data into a system. In a CoT system where the reliability of the sensed data is crucial, injecting any false data can compromise the whole security of the system 0.

- **Weak cryptographic protocols:**

IoT devices have limited computational capabilities and power sources. This may influence the developer of a CoT system to employ weaker cryptographic mechanisms which require less computational power because standard cryptographic mechanisms are computationally intensive. Such weak cryptographic mechanisms might be easily breakable which can threaten the security of the whole system 0.

- **Data loss and leakage:**

According to a report of Cloud Security Alliance, data loss

could occur for many reasons. For instance, accidental deletion of data by the service provider, physical catastrophe like earthquake and fire or lack of disaster recovery as well as unreliability of the data centre 0. Similarly, data leakage could occur for other reasons such as exploitations by malicious insiders, weak access control mechanisms, lack of powerful encryption algorithms, etc. In addition, data leakage could occur in a CoT system where computing and hardware resources are shared.

- **Data breaches:**

Data breaches are situations in which data is accessed by an unauthorized or illegitimate entity which can be a person, application or service. Data breaches can occur at any layer in a system as well as at endpoints. Data breaches might occur from outside the system initiated by an outside attacker or inside within the system by a malicious insider 0.

- **Data sensitivity:**

Some sensor data (e.g., medical data, financial data, etc.) are more sensitive in nature than others. Generally, sensitive data imposes greater risks for invading privacy and breaching security of a user. Therefore, more care should be taken in determining the sensitivity of such data as well as in storing and utilising such data.

A-4.   Service provisioning threats

Threats which are related to different services within a CoT system belong to this sub-category. The identified threats and attacks are presented below.

- **Unidentified and unauthorised access:**

CoT-enabled services that deal with sensor data must ensure

that unidentified and unauthorised users cannot access data. Otherwise an attacker can easily gain access to sensitive sensor data and can the results of analysis of the data.

- **Escalation of privileges:**

Inadequate deployment of authorisation infrastructures while providing CoT-enabled services may allow a user or attacker to gain unprivileged accesses to sensitive sensor data and their resulting analysis.

- **Identity theft:**

In an identity theft attack, an attacker tries to gain illegal access to restricted services and resources by pretending to be a valid user of the system using key pieces of personal information of a valid user. The user concerned may also suffer from losses and can be held accountable for the attacker's actions 0.

- **Service hijacking:**

In a service hijacking attack, a user, while trying to access a valid service, is redirected to an illegal service controlled by the attacker. The user is then tricked to reveal personal information which could be exploited for other types of attacks such as identity theft, manipulating user data, spying user activities and transactions, returning fake information, etc. There are different ways such an attack can be launched such as phishing attacks, social engineering attacks, using exploited software, etc. 0.

- **Insecure interfaces and API:**

Insecure interfaces and application programing interfaces are amongst the top threats in Cloud Computing. This is because a set of application programing interfaces (API) is widely distributed by Cloud service providers to allow their consumers to retrieve data and services 0. Attackers can exploit weaknesses in interfaces which are not properly protected 0.

- **Compromising management interface**

CoT systems are generally managed over the Internet using Client management interfaces. Protecting such management interfaces is crucial as an insecure interface could be exploited to launch fraudulent services that utilise legitimate data sensed and produced by IoT devices 0.

A-5.   Other threats

In this sub-category, miscellaneous threats which are not related to any of the previous sub-categories are listed.

- **Malicious insiders:**

A malicious insider is a risk to any organisation. A malicious insider can be a current or previous employee of an organisation and may have authorised access to an organisation's systems or have access to potential sensitive data. The insider may exploit such privilages to abuse the data as well as services. It is important for organisations to understand what providers are doing to identify and protect against the malicious insiders 0, 0.

- **Shared technology issues:**

In a CoT system, shared resources might by utilised via virtualisation and multi-tenancy architecture which allows many clients to share the same application instance. Sharing an application amongst many users in this way may permit a user to access the virtual machine (VM) of another user. Alternatively, any vulnerabilities of the VM Monitor (VMM) could be exploited by a malicious user to gain access to another user's VM.

It is difficult to ensure strong isolation between tenants when the IaaS is delivered through multi-tenancy architecture 0.

- **Abusing Cloud Computing:**

A big benefit of Cloud Computing is that it allows an organisation of any size an access to large volumes of computing power which would be impossible for such organisation otherwise 0, 0. However, there is a drawback that comes with this opportunity as this allows attackers to abuse such computing power to launch different types of attacks. For example, an attacker can rent significant computering power and use it to launch DoS attacks against another cloud service provider. The Cloud Security Alliance considers this to be a major security threat to Cloud Computing which is applicable for CoT 0.

B. Privacy Threats

The following threats can be exploited to invade the privacy of different entities in a CoT system. Since the matter of privacy mainly affects the users of the system, these threats are more focused on how they can breach or invade the privacy of the users within the corresponding system.

- **Unnoticed capture & identitification:**

An IoT device deployed in an environment to collect data about users (e.g., a miniscule size camera deployed in the entry point of a building capturing facial images of every person entering the building) can be discrete in nature. Such data can be captured without the knowledge or consent of the users and then can be analysed and abused to identify users and invade their privacy.

- **IoT data inaccesibility:**

Data captured by different IoT devices in an environment and then uploaded to the cloud providers may not be accessible to the user whose data has been captured. Without having any access, the user will have no knowledge how and what data have been collected.

- **Lack of control and transparency:**

Once collected data about users are uploaded to the cloud, the users have limited control over them and they may be unaware how such data are being used or abused by the data collector. The very nature of the ubiquituous sensing process which may capture data in a continuous stream makes it very difficult for the users to express their consent explicitly regarding their acceptance or denial regarding data collection or what to do with the data collected while they are processed, analysed, presented and shared in a system. Without such controls it becomes difficult to create access control rules in a system to protect the privacy of the system.

- **Loss of governance:**

Loss of governance is considered one of the most important classes of cloud specific threats. In using the cloud infrastructures, the customers essentially give control to the cloud provider which might impact their privacy.

- **Profiling and tracking:**

Where data collected in a variety of different environments can be identified with particular users, it may be possible to create profiles of users who may then be tracked her across multiple domains without their knowledge.

- **Unforseen inference:**

Data collected from different sensors can be analysed comprehensively using extensive computing power offered by

CoT. Wth this extensive analysis comes a risk of unforeseen inference regarding users which otherwise would not be possible. The knowledge gained from such inference could be exploited to invade the privacy of users.

- **Unatuhorised disclosure:**

Data collected by different IoT devices will be disclosed or uploaded to a CoT system for storage and further analysis. It is extremely difficult to collect consent or even notify users regarding data collection, storage and analysis procedure in this manner which might result in unauthorised disclosure of (sensitive) data.

Conclusions

Cloud Computing offers a new processing model that increases efficiency and provides on-demand access to a shared pool of resources with minimum management effort in a considerable lower cost. On the otherhand, IoT provides a platform to inter-connect an array of devieces and enable them to interact with each other in a seamless manner. Combining these two different technical paradigms to enable the storage and processing of unprecedented amount of data generated by different IoT devices seems to be the next step in the technological evolution which can be leveraged to offer innovative cloud-based online services. The model of CoT is the realisation of this idea and has emerged as a platform to allow intelligent usage of applications and sensors as services based on real time data from these IoT devices. Being a new technical paradigm, there are security and privacy issues involved in CoT which need to be addressed before its full potential can be harnessed. The main goal of this chapter is to present an overview of different security and privacy threats within the domain of CoT.

Toward this goal, a threat model has been presented consisting of a model of an adversary along with a list of assests. Then, a taxonomy containing different security and privacy threats has been presented. Finally, each identified threat has been briefly discussed. Many of these threats emerge due to the intrinsic nature of Co|T where data generated by diffrerent IoT devices may involve a human. Depending on the data generating device, such personal data can be extremely sensitive in nature. Such data may be collected ubiquitiously in an automatic fashion; where the respective person may be unaware of the data collection procedure. This makes it difficult for any person to provide explicit consent during data collection. Furthermore, data will be collected in different environments managed by different authorities. Managing such personal data generated in largle volumes in different environments will be extremely challenging for any user. All these properties of CoT will challenge the existing practices of security and privacy preserving mechanisms that deal with personal data.

Similarly, organisations responsible for the deployment of CoT systems will face challenges with respect to maintaining security and preserving privacy of the colleced data. Novel secure mechanisms will need to be deployed to collect IoT data from environments, transmit such data via the Internet and finally store and analyse them in a cloud. Moreover, novel access control mechanisms will need to be deployed to ensure that any single person can is informed what data is being collected and has the provision for the user to express explicit consent. In addition, the organisations will also need to deploy novel services to allow users to manage their collected data in a secure, usable and privacy-friendly way.

One way to design and develop a secure and privacy-friendly CoT system is to ensure that the underlying threats are addressed. This chapter lays out a foundation toward that goal by identifiying the threats within a CoT system. The next challenge will be to produce a comprehensive list of mitigation strategies which can be utilised to formulate security and privacy requirements. Fulfilling these requirements will ensure that the threats are mitigated, however, the ultimate challenge is to design a system that can satisfy all these requirements. These challenges open up an exiciting oportunitiy of extensive research in the coming years.

REFERENCES

[1] Kim, S.H. and Kim, D. (2015). Enabling Multi-tenancy via Middleware-level Virtualization with Organization Management in the Cloud of Things. Services Computing, IEEE Transactions on, 8(6), pp.971-984.

[2] Alohali, B., Merabti, M., & Kifayat, K. (2014). A Secure Scheme for a Smart House Based on Cloud of Things (CoT), 115–120.

[3] Aazam, M., Khan, I., Alsaffar, A.A., and Huh, E.N.(2014). Cloud of Things: Integrating Internet of Things and Cloud Computing and the issues involved. In-Applied Sciences and Technology (IBCAST), 2014 11th International Bhurban Conference on (pp. 414-419). IEEE.

[4] Buyya, R., Yeo, C., & Venugopal, S. (2008). Market-oriented Cloud Computing: Vision, hype, and reality for

delivering it services as computing utilities. CoRR, (abs/0808.3558).

[5] Vaquero, L. M., Rodero-Merino, L., Caceres, J., & Lindner, M. (2008). A break in the clouds: towards a cloud definition. ACM SIGCOMM Computer Communication Review, 39(1), 50-55.

[6] Mell, P., & Grance, T. (2011). The NIST Definition of Cloud Computing Recommendations of the National Institute of Standards and Technology. Nist Special Publication, 145, 7. doi:10.1136/emj.2010.096966

[7] Armbrust, M., Fox, A., Griffith, R., Joseph, A., & RH. (2009). Above the clouds: A Berkeley view of Cloud Computing. University of California, Berkeley, Tech. Rep. UCB, 07–013.[Online] Article available at < www.eecs.berkeley.edu/Pubs/TechRpts/2009/EECS-2009-28.pdf > [Accessed: 27st April 2016] .

[8] Ashton, K., (2009.) That 'internet of things' thing. RFiD Journal, 22(7), pp.97-114.

[9] EPOSS.(2008). Internet of Things in 2020, Roadmap for the Future, Version 1.1, 27 [online] Avilable online at<http://www.smart-systems integration.org/public/documents/publications/Internet-of-Things_in_2020_EC-EPoSS_Workshop_Report_2008_v3.pdf> [accessed on: 28/04/2016].

[10] Dunkels, A., Vasseur J.(2008). IP for Smart Objects, Internet Protocol for Smart Objects (IPSO) Alliance, White Paper #1, [online] availlable at <

http://dunkels.com/adam/dunkels08ipso.pdf > [accessed on 28/04.2016]

[11] Guillemin, P., and Friess, P. (2009) "Internet of things strategic research roadmap," The Cluster of European Research Projects, Tech. Rep. [online] avilable at <http://www.internet-of-things-research.eu/pdf/IoTClusterStrategicResearchAgenda2009.pdf [Accessed on: 2016-04- 26].

[12] Atzori, L., Iera, A. and Morabito, G.,. (2010). The internet of things: A survey. Computer networks, 54(15), pp.2787-2805.

[13] La Diega, G.N. and Walden, I., (2016). Contracting for the 'Internet of Things': Looking into the Nest. Queen Mary School of Law Legal Studies Research Paper No. 219/2016.

[14] Khan, R., Khan, S. (2012). Future Internet: The Internet of Things Architecture, Possible Applications and Key Challenges, in the proceedings of 10th International Conference on Frontiers of Information Technology, Islamabad, Pakistan, 17-19 D.

[15] Wu, M. 2012. Research on the architecture of Internet of things, in the proceedings of 3rd International Conference on Advanced Computer Theory and Engineering, 20-22, Beijing, China.

[16] Gubbi, J., Buyya, R., Marusic, S. and Palaniswami, M., (2013). Internet of Things (IoT): A vision, architectural elements, and future directions. Future Generation Computer Systems, 29(7), pp.1645-1660.

[17] Tei, K. and Gurgen, L., 2014, March. Clout: Cloud of things for empowering the citizen clout in smart cities. In Internet of Things (WF-IoT), 2014 IEEE World Forum on (pp. 369-370). IEEE.

[18] Bhattasali, T., Chaki, R. and Chaki, N. (2013). Secure and trusted cloud of things. In India Conference (INDICON), 2013 Annual IEEE (pp. 1-6). IEEE.

[19] Tanganelli, G., Vallati, C. and Mingozzi, E. (2014). Energy-efficient QoS-aware service allocation for the cloud of things. In Cloud Computing Technology and Science (CloudCom) 2014 IEEE 6th International Conference on (pp. 787-792). IEEE.

[20] Distefano, S., Merlino, G., and Puliafito, A. (2012). Enabling the cloud of things. In Innovative Mobile and Internet Services in Ubiquitous Computing (IMIS), 2012 Sixth International Conference on (pp. 858-863). IEEE.

[21] Myagmar, S., Lee, A. J., & Yurcik, W. (2005, August). Threat modeling as a basis for security requirements. In Symposium on requirements engineering for information security (SREIS) (Vol. 2005, pp. 1-8).

[22] Desmet, L., Jacobs, B., Piessens, F., & Joosen, W. (2005). Threat modelling for web services based web applications. In Communications and multimedia security (pp. 131-144). Springer US.

[23] De Cock, D., Wouters, K., Schellekens, D., Singelee, D., & Preneel, B. (2005). Threat modelling for security

tokens in web applications. In Communications and Multimedia Security (pp. 183-193). Springer US.

[24] Dolev, D., & Yao, A. C. (1983). On the security of public key protocols.Information Theory, IEEE Transactions on, 29(2), 198-208.

[25] Goldreich, O. (2009). Foundations of cryptography: volume 2, basic applications. Cambridge university press.

[26] Paverd, A. J., Martin, A., & Brown, I. Modelling and Automatically Analysing Privacy Properties for Honest-but-Curious Adversaries. Tech. Rep., 2014.[Online]. Available: https://www. cs. ox. ac. uk/people/andrew. paverd/casper/casper-privacy-report. pdf.

[27] Alotaibi, K. H. (2015). Threat in Cloud- Denial of Service (DoS) and Distributed Denial of Service (DDoS) Attack , and Security Measures. Journal of Emerging Trends in Computing and Information Sciences, 6(5), 241–244.

[28] Jensen, M., Schwenk, J., Gruschka, N., & Iacono, L. Lo. (2009). On technical security issues in Cloud Computing. In CLOUD 2009 - 2009 IEEE International Conference on Cloud Computing (pp. 109–116).

[29] Mahalle, P., & Anggorojati, B. (2012). Identity establishment and capability based access control (IECAC) scheme for Internet of Things. … (WPMC), 2012   15th   …,   184–188.   Retrieved   from http://ieeexplore.ieee.org/xpls/abs_all.jsp?arnumber=639 8758

[30] Li, Y., & Teraoka, F. (2012). Privacy protection for low-cost RFID tags in IoT systems. Proceedings of the 7th International Conference on Future Internet Technologies - CFI '12, 60. http://doi.org/10.1145/2377310.2377335

[31] Vidalis, S., & Olga, A. (2014). Assessing Identity Theft in the Internet of Things. IT CoNvergence PRActice (INPRA), 2(1), 15–21.

[32] Stojmenovic, I., & Wen, S. (2014). The Fog Computing Paradigm: Scenarios and Security Issues. Proceedings of the 2014 Federated Conference on Computer Science and Information Systems, 2, 1–8. http://doi.org/10.15439/2014F503

[33] Liu, J., Xiao, Y., & Chen, C. L. P. (2012). Authentication and Access Control in the Internet of Things. 2012 32nd International Conference on Distributed Computing Systems Workshops, 588–592. http://doi.org/10.1109/ICDCSW.2012.23

[34] Gope, P., & Hwang, T. (2015). Untraceable Sensor Movement in Distributed IoT Infrastructure. IEEE Sensors Journal, 15(9), 5340–5348. http://doi.org/10.1109/JSEN.2015.2441113

[35] Roman, R., Zhou, J., & Lopez, J. (2013). On the features and challenges of security and privacy in distributed internet of things. Computer Networks, 57(10), 2266–2279. http://doi.org/10.1016/j.comnet.2012.12.018

[36] Zhao, K., & Ge, L. (2013). A survey on the internet of things security. Proceedings - 9th International Conference on Computational Intelligence and Security, CIS 2013, 663–667. http://doi.org/10.1109/CIS.2013.145

[37] Babar, S., Stango, A., Prasad, N., Sen, J., & Prasad, R. (2011, February). Proposed embedded security framework for internet of things (iot). InWireless Communication, Vehicular Technology, Information Theory and Aerospace & Electronic Systems Technology (Wireless VITAE), 2011 2nd International Conference on (pp. 1-5). IEEE.

[38] Verbauwhede, I. (2010). Secure Integrated Circuits and Systems, 250. http://doi.org/10.1007/978-0-387-71829-3

[39] Babar, S., Mahalle, P., Stango, A., Prasad, N., & Prasad, R. (2010). Proposed security model and threat taxonomy for the internet of things (IoT). In Recent Trends in Network Security and Applications (pp. 420-429). Springer Berlin Heidelberg.

[40] Mahalle, P., Babar, S., Prasad, N. R., & Prasad, R. (2010). Identity management framework towards internet of things (IoT): Roadmap and key challenges. In Recent Trends in Network Security and Applications (pp. 430-439). Springer Berlin Heidelberg.

[41] Komninos, N., Philippou, E., & Pitsillides, A. (2014). Survey in smart grid and smart home security: issues, challenges and countermeasures.Communications Surveys & Tutorials, IEEE, 16(4), 1933-1954.

[42]Clark, J. A., Murdoch, J., McDermid, J. A., Sen, S., Chivers, H., Worthington, O., & Rohatgi, P. (2007,

September). Threat modelling for mobile ad hoc and sensor networks. In Annual Conference of ITA (pp. 25-27).

[43] Cloud Security Alliance. (2013). The Notorious Nine. Cloud Computing Top Threats in 2013. Security, (February), 1–14.

[44] Subashini, S., & Kavitha, V. (2011). A survey on security issues in service delivery models of Cloud Computing. Journal of Network and Computer Applications, 34(1), 1–11. http://doi.org/10.1016/j.jnca.2010.07.006

[45] Modi, C., Patel, D., Borisaniya, B., Patel, A., & Rajarajan, M. (2013a). A survey on security issues and solutions at different layers of Cloud Computing. Journal of Supercomputing, 63(2), 561–592. http://doi.org/10.1007/s11227-012-0831-5

[46] Pearson, S., & Benameur, A. (2010). Privacy, Security and Trust Issues Arising from Cloud Computing. 2010 IEEE Second International Conference on Cloud Computing Technology and Science, 693–702. http://doi.org/10.1109/CloudCom.2010.66

[47] Babu, S., Ph, C., Bansal, V., & Telang, P. (2010). Cisco: Top 10 Cloud Risks That Will Keep You Awake at Night, 1–35. Retrieved from https://www.owasp.org/images/4/47/Cloud-Top10-Security-Risks.pdf

[48] Catteddu, D., & Hogben, G. (2009). The European Network and Information Security Agency (ENISA) is

an EU agency created to advance This work takes place in the context of ENISA ' s Emerging and Future Risk programme . C ONTACT DETAILS : This report has been edited by. Computing, 72(1), 2009–2013. http://doi.org/10.1007/978-3-642-16120-9_9

[49] Kajiyama, T. (2012). Cloud Computing Security: How Risks and Threats are Affecting Cloud Adoption Decisions ***, 105. Retrieved from http://sdsu-dspace.calstate.edu/handle/10211.10/3522

[50] Hashizume, K., Rosado, D., Fernández-Medina, E., & Fernandez, E. (2013). An analysis of security issues for Cloud Computing. Journal of Internet Services and Applications, 4(5), 1–13. http://doi.org/10.1186/1869-0238-4-5

[51] Bensoussan, A. (2011). Impact of Security Risks on Cloud Computing Adoption, 670–674.

CHAPTER SEVEN

SMART HOMES BASED ON SMART CITIES' DESIGN PATTERNS

Basman Alhafidh [a b]   Hayder Khzaali [a c]

Ahmed Mahmood [a d]   William Allen [e]

a: Department of Electrical and Computer Engineering, Florida Institute of Technology, FL, USA.

b: Department of Computer Engineering, University of Mosul, Mosul, Iraq

c: Department of Computer Engineering and Info. Technology, University of Technology, Iraq.

d: Department of Electrical Engineering, University of Mosul, Mosul, Iraq.

e: School of Computing, Florida Inistitute of Technology, Melbourne, FL, USA.

Abstract

The new developing technology that we can see every day in computers, networks, and control systems is making a huge change in our daily life, especially with exciting Internet Services. Therefore, the Internet of Things is playing an important role in managing home sensors, actuators, and devices to end up with a smart city. This chapter, firstly, presents a review of smart cities for the Internet of Things about maximizing the efficiency of distribution and consumption of energy on the one side and a vision for smart cities in the future on the other.

Secondly, it presents the design for a smart home and ends up with a proposed system, as a case study, to enhance the design of the smart home environment. This introduces a novel approach by using a smart agent for each subsystem, a storage agent with the cloud for backup data with an enterprise analysis, and the central intelligent agent (Brain) that connects all the components of the proposed system in one unique design approach. The presented design can learn a user's behaviors and then independently perform tasks for the user, often anticipating the user's needs. Additionally, it presents an integrated paradigm for a green architecture design approach. Finally, the study shows an integrated platform for the Internet of Things paradigm to support wire/wireless sensing systems and information processing efficiently.

Keywords: Smart City, Smart Home, Intelligent Agent, Self-Adaptive System.

INTRODUCTION

For any city development around the world, the Internet of Things is the best technology key for integrating many points, such as big data and analytics, mobility, and cloud services. It attains sustainable economic developments and a high quality of life. Smart cities, with intelligent management systems, offer the solution to current challenges in addition to presenting new services.

There is no agreement among experts on a single definition for the concept" Smart City." However, most recent identifications share the same common theme, which is using up to-date technology to take advantage of whatever public resources are available to enhance the quality and quantity of services provided to them. Some of these definitions are the following:

• "Smart Cities are a future reality for municipalities around the world. These cities will use the power of ubiquitous communication networks, highly distributed wireless sensor technology, and intelligent management systems to solve current and future challenges and create exciting new services" [1].

• "Smart city is a city that uses digital technologies to enhance the quality of life and standard of living of its citizens" [2].

• "Smart city is a powerful paradigm that applies the most advanced communication technologies to urban environments, with the final aim of enhancing the quality of life in cities and providing a wide set of value-added services to both citizens and administration" [3].

• "Smart city is a city that has deployed, or is currently piloting the integration of, ICT solutions across three or more of mobility and transport, energy and sustainability, physical infrastructure, governance, and safety and security city-functions in order to improve efficiency, manage complexity and enhance citizen quality of life, leading to a sustainable improvement in city operation" [4].

• "A city connecting the physical infrastructure, the IT infrastructure, the social infrastructure, and the business infrastructure to leverage the collective intelligence of the city" [5].

Figure 1 illustrates the number of cities involved in smart city development and their associated investment in this field, taking into account important functions that include: mobility and transport, governance, energy and sustainability, safety, security, and physical infrastructure [6].

SMART CITY SERVICES

To appreciate the necessity for smart cities, we shed some light on a few of its services:

1) Structural Health of Important Buildings

Some buildings, such as historical buildings, need special care and continued attention to identify any vulnerable or weak points that might occur in its structure due to external events that affect these buildings over time to keep them in good physical condition. In a smart city, this issue has been solved by deploying specific types of sensors, like deformation or vibration sensors, these are connected wirelessly to a sink node, which reports the data to central management.

Then, a team can be dispatched to repair any reported issues instead of carrying out periodic checking, which wastes both money and human resources [6], [7].

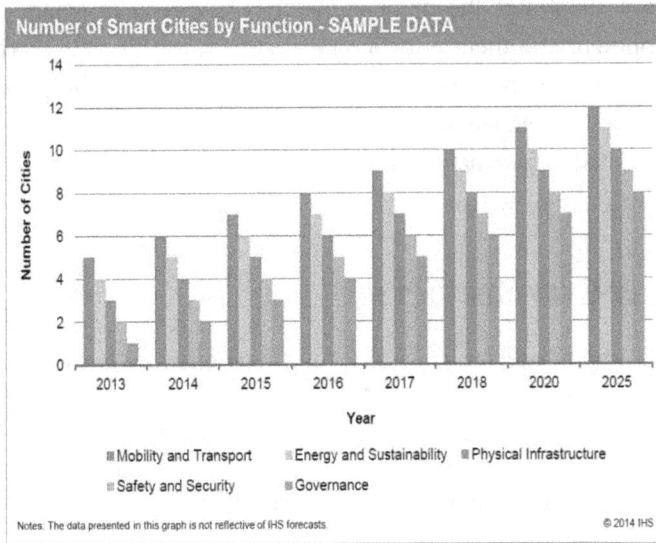

Figure (1). Number of smart cities by function [6]

2) Waste Management

One of the most important issues in each and every country is the waste management problem because of two existing facts. One is the storage of garbage in any specific location and the other is related to the operational cost. It is known that the same garbage truck returns to the same place two or three times a day.

An effective approach to solve both problems in a smart city is by putting special sensors that measure the level of garbage in the container and send that information to central management. In turn, management notifies the garbage truck driver in charge of that area to collect it and presents him with the shortest and most effective route to the destination.

This process will cause significant economic savings for the country [7].

3) Air Quality

Smart city services in this field provide a means to monitor air quality in crowded areas, along fitness trails, in parks and the like [7]. People can now know which path is healthier to select for outdoor activities. For example, this can be implemented by deploying air pollution sensors that sense the weather and report the measured data to the public.

4) Traffic Congestion

Although the old method for traffic monitoring, using camera-based technology, is currently being used in many countries, using low-powered sensors with a GPS connection capability and wide-range communications in modern vehicles can provide a denser source of information by connecting these sensors to traffic sink nodes [8].

5) Revealing new information

With the deployment of many types of sensors or information gathering equipment, city leaders will be able to uncover information within their cities as never before. For example, in San Francisco, the police officials assumed that the crime rate was very low because they believed that residents would call to report a gunshot 80% of the time. However, many acoustic sensors that were modified to detect gunshots were deployed after that. After reading the measurements from the sensors, the police officials were shocked to discover that apparently residents called 911 when they heard a gunshot only 10% of the time. People can also discover hidden facts at home;

for example, residents pay their electric bills every month without knowing the details about the electric expenditure of each and every electric device in their homes. This problem can be handled by the use of smart plugs that have the capability of accurate profiling or monitoring, thus providing details about the electric power consumed at the appliance level [9].

6) Smart lighting

For indoor lighting, energy consumption of light can be decreased by 70% with the installation of smart lighting systems that depend on the existence of daylight and the use of technology to blend electrical light with daylight. However, with outdoor lighting, many cities around the world still use the antiquated or very old design street lights, which waste electric energy and money as well as cause pollution. In a smart city, this is not the case. Each lamppost's out-dated electrical light is replaced with an efficient LED light and not only that, it is connected to an access point to form a mesh network. In this approach, we can cover major parts of the city with Internet services in addition to having the ability to control the electric power and brightness of the color of every lamppost. These improvements will make a drastic change to the economy [9].

STAGES OF SMART CITY DEVELOPMENT

The basic structure is to use information and communication to enhance the citizens' quality of life. Although a fully integrated Smart city is still a future goal for most advanced countries, many steps have been taken to reach this vision. Issues and problems for most city leaders are nearly the same; for example, a premature technology infrastructure compared to other departmental or cultural factors can affect the progress of smart city development.

Therefore, a maturity model with several stages has been created (by IDC government insights) for a smart city to help city leaders determine where they are and what they should do next. This maturity model is illustrated in Figure 2.

Stages in Smart City Development

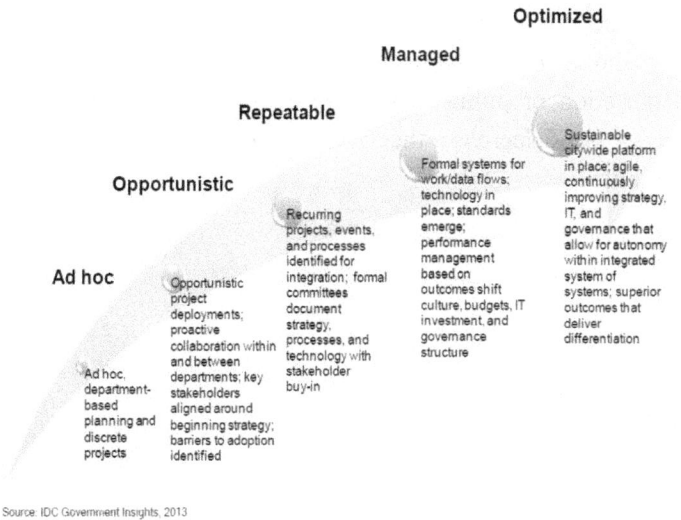

Source: IDC Government Insights, 2013

Figure (2). Stages in smart city development [1]

There are many reasons behind investment in smart city projects. However, these reasons differ from city to city depending on where the city is located in the smart city maturity model and what priorities city leaders prefer to achieve first. A general theme is to reduce the operational cost, but there are other factors that are different for each city. Traffic congestion may drive city leaders to invest in real time traffic information systems by developing specific types of sensors that have certain capabilities to solve this issue. Alternatively, if citizens have complained about the availability of parking spaces, then that will drive city leaders to invest in that area. Figure 3 illustrates different sectors being studied and analyzed at the global level. It is clear that there is no equal maturity among the different departments, and sometimes, even inside the same division. For example, in the ad-hoc stage, both public safety and works are found, but for the data practice area, the transportation sectors are located between the ad-hoc and opportunistic stage. At the same time, public safety shows more maturity in the process sector than does transportation or public works. As a consequence, this uneven progress may decrease the overall growth of the smart city development.

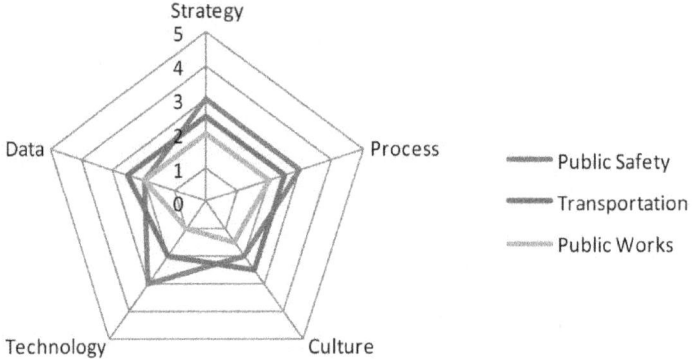

Figure (3). City baseline department: IDC Government insights [1]

THE SMART CITY VISION

Achieving a smart city vision requires answering a few questions first:

1) How do citizens appreciate the new ideas that will make a significant impact on their life?

This is a crucial factor that often leads to either a potential failure or success for city leaders. The whole smart city dream will diminish unless these people have the will to experience new ideas, are ready to cooperate with their government departments and stakeholders, and embrace the process for building upon failures and success until finding what works well.

2) How does one find a sponsor for this long term project vision?

The required budget for this long-standing project is tremendous and it is not sufficient or convenient to rely only on what the central government provides.

The main reason for that is not to allow the governments to make the decision for what areas should be invested in, based only on their opinions and what they see fits their needs and ignoring the cities necessities and visions. When that happens, cities will no longer have the right to choose what domains are best suited for their interests.

3) What should city leaders consider first when choosing types of services?

Cities need leaders who can figure out or develop a digital master plan that covers the whole city. Building this plan is essential and represents the building block for the smart city project and the first services that leaders should think of.

This is not a ten- or fifteen-year budget, but rather a long-term plan that requires a thorough view of the key challenges that face the city and what should be done to solve these issues [1].

SMART HOME

A smart city includes many components, including smart grids, smart homes, and smart buildings. A smart home as a constitutive part of smart city structure is discussed in this section.

A smart home is any construction that has a certain design supported with several technologies to enhance or maximize its functionalities and services for its occupants, while extending the life of its physical structure and decreasing the operational cost at the same time. A general definition of a smart home can be represented by a home that has different subsystems that are connected together and that use their stakeholder's information and database management to implement the delegated goals. This platform makes use of the connected network cabled to a computer to manage the data. In addition, new smart homes try to serve the stakeholder with many control capabilities from inside or outside the home. Such services include exchanging the information smoothly, mediating stakeholder lifestyle, arranging the work schedule in the calendar, managing the security, and saving money through optimizing energy consumption [10].

We can say that the new smart homes are a typical module for the Internet of Things, which presents a very interesting field for both the academic and industrial framework. Figure 4, shows that the Internet of Things (IoT) has four application domains, which are classified according to the type of network coverage and availability, scalability, and user interactions. It also shows that the home environment is one of the IoT classified applications, which actually represents a typical example of IoT technology. The figure shows how much the domains are interconnected using the Internet, which enables the sharing of data between a wide range of providers in a smooth way to create different business opportunities [10].

Basically, the IoT framework consists of many wireless sensors and actuators in the smart field. These nodes play an important role in controlling and optimizing intelligent buildings and even smart greenhouses. In other words, using distributed nodes of sensors and actuators in the home environment in addition to an optimized data management system will be carried out under the scope of IoT principles. As a result, such a system can precisely monitor stakeholder behavior and evaluate an action, trying to anticipate the user's behavior in the future. This could include monitoring elderly people in a healthy environment [11].

Many of the scientific community do not give a precise definition of a smart home, since some authors are not satisfied about which elements can be part of a smart home, so what can be used to easily classify this term from another.

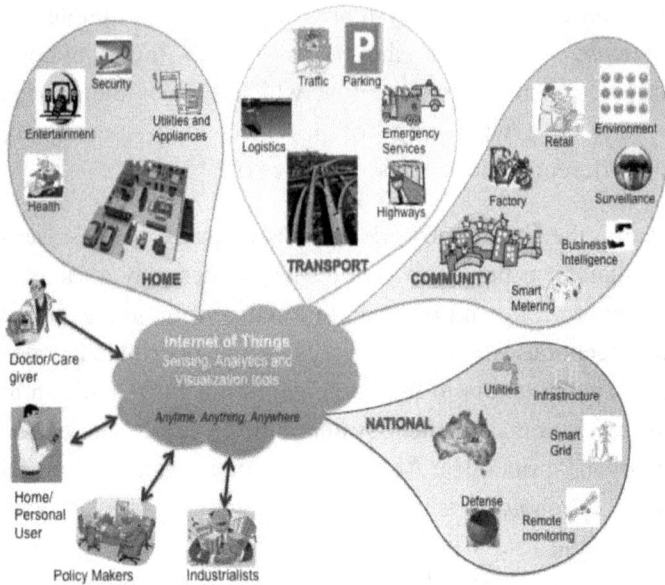

Figure (4): Internet of Things Schematic showing the end user
and application area based on data [10]

Silva et al. [12] define smart homes as a "home-like environment
that possesses ambient intelligence and automatic control." This
means that a home has the ability to react with the stakeholder
and to give different types of services classified into the
following four types: security based services, energy efficiency
based, health-care based, and multimedia and entertainment
based smart homes [13].

In [13], the researchers present another idea about smart homes
as "The backbone which will enable the management and the
control of different areas of a residence, binding four pillars of
human livelihood inside a house: comfort and welfare; physical

integrity and facility safety; rational management of domestic energy equipment; and the possibility to provide healthcare services to its inhabitants." The authors add that the important thing in this term is communication. Communication should be established in an easy, cheap, and trusted way with a robust communication structure. So, a further definition for smart homes is that a "Smart Home can be defined as a concentrator and disseminator of information and service." Both information and services try to provide a gateway to outdoor world services such as a smart grid side by side with the ability to provide indoor services or local functionalities to improve the level of quality of life. Such a gateway will open a wider window of services by sharing the managed local information with national services like those found in smart cities.

SMART HOME DESIGN

All the interconnection and data exchanging between home devices refers to an ecosystem of devices. These heterogeneous devices include but are not limited to home appliances, sensors, actuators, and all the subsystems and gateways in the interconnected system. Much research is concentrated on how to optimize the power consumption at home. Many of the industry solutions focus on connecting the smart home to a smart grid like the Energy@home work that gives a good example of interfacing the home appliances with a smart grid. Such a system allows the devices in the environment operate on an electrical outlet when the consumption of their energy is low.

On the other hand, much research focuses on how to develop the smart gateway that interface all nodes in the environment to provide more efficient and entertainment services, but many challenges appear when trying to eliminate the large number of smart boxes inside the home environment [14].

From the technical point of view, both wired and wireless communication methods are presented in a home network system. Year by year, the wireless technique is growing faster than the wired one, since wireless devices have more flexibility when a customer or a specialist tries to install the nodes inside homes. In addition, the power outlets inside a home can supply limited power resources for wireless nodes. It is very common to see more than one wireless technology in a given home. These technologies include Bluetooth, ZigBee, Wi-Fi, WiMAX, and Z-wave. Table 1 shows a comparison between many wireless communication technologies. Many debates mention that the ZigBee technology is the best and most common technology to use in home automation, while others recommend Z-wave to get a higher level of performance [15].

It would be very interesting to have a technology that has both performance and affordability with a high level index. Unfortunately, there is no technology present at a high level for both indexes. We can see clearly in Figure 5 that X10 is the best choice between the competitors when we talk about affordability, but at the same time, it has the lowest performance level. On the other hand, Z-Wave presents the best performance index among all technologies with a medium affordability index. This knowledge makes Z-Wave the best choice for stakeholders when they intend to build a smart environment. However, the unique features of ZigBee, which is very open, makes the ZigBee slightly more preferred than Z-Wave [16].

Information Processing

In the smart home monitoring system, there are key functionalities and basic properties for the information processing system related to this monitoring environment. These keys include the flexibility, robustness, the compatibility of the sub-system, and real-time processing of data. Basically, any automation system may contain different subsystems for multiple purposes. So, the compatibility requirement is a primary demand in this case. Otherwise, it would be very complex for each subsystem to communicate with other subsystems. As a result, it would be hard to analyze human activities [15].

Smart Home Architecture

From the information processing paragraph, we can say that there is a difficulty in integration between many heterogeneous devices from different subsystems in a given environment. So it is essential to integrate between all the subsystems to easily manage the data and get the information by analyzing the data to supply the smart environment with extraordinary services. One of the solutions to present such a system is to divide the system into four elements: business platform, smart gateway, wireless nodes, and mobile customer device to monitor the environment. These elements are concurrently located in the business platform layer, device access layer, device layer, and application layer, as shown in Figure 6.

Table (1). Comparison of the Popular Home Automation Technologies [17]

Performance Factors	Z-Wave	ZigBee	X10	INSTEON	EnOcean
Released (Year)	2001	2004	1975	2005	2008
Inventor	ZenSys Corp.	ZigBee Alliance	Pico Electronics	Smartlabs Inc.	EnOcean GmbH
Standardization	Proprietary	IEEE 802.15.4	Proprietary	Proprietary	Proprietary
Primary Markets	Home Automation	Industrial Automation, Research, Home Automation, Telecommunications, Healthcare	Home Automation	Home Automation	Industrial Automation, Home Automation
Communication Mode	RF	RF	RF; Power Line	RF; Power Line	RF
System-On-Chip Solution	Yes	Yes	Yes	Yes	Yes
Encryption	128-bit AES	128-bit AES	No	No	ARC4/AES
Energy Usage	High (1)	Medium (2)	High (1)	High (1)	Nil (3)
Data Rate	~40 kbps (3)	>20 kbps (3)	20-200 bps (1)	~2000 bps (1)	125 kbps (3)
Two-way Communication	Yes (3)	Yes (3)	No (0)	Yes (3)	Yes (3)
Transmission Range	~120m (3)	~60m (2)	~30m 2)	~120m (3)	> 20m (2)
Inter-brand Operability	High (3)	Medium (2)	Low (1)	Medium (2)	Medium (2)
Number of Certified Devices	>600 (3)	<500 (2)	>500 (3)	<500 (2)	>600 (3)
Ability to work as Repeaters	Yes (3)	Yes (3)	No (0)	Yes (3)	No (0)
Ease of Installation	Easy (3)	Medium (2)	Difficult (1)	Easy (3)	Medium (2)
Performance Index	0.916	0.792	0.375	0.75	0.75
Affordability Index	0.34	0.212	1.00	0.362	0.46

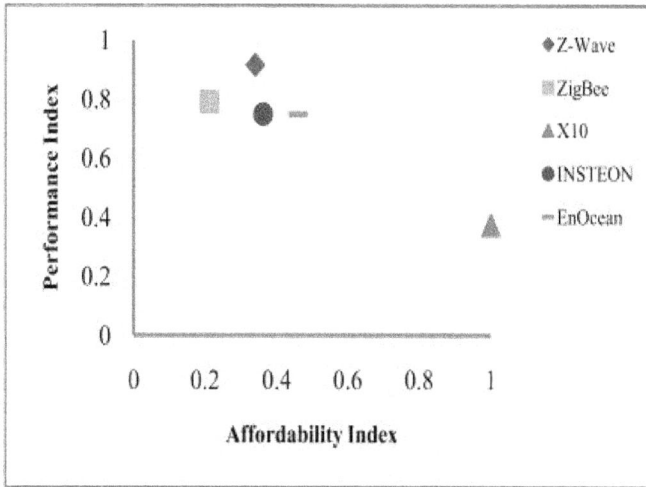

Figure (5). Performance vs Affordability of the Popular Home
Automation Technologies [16]

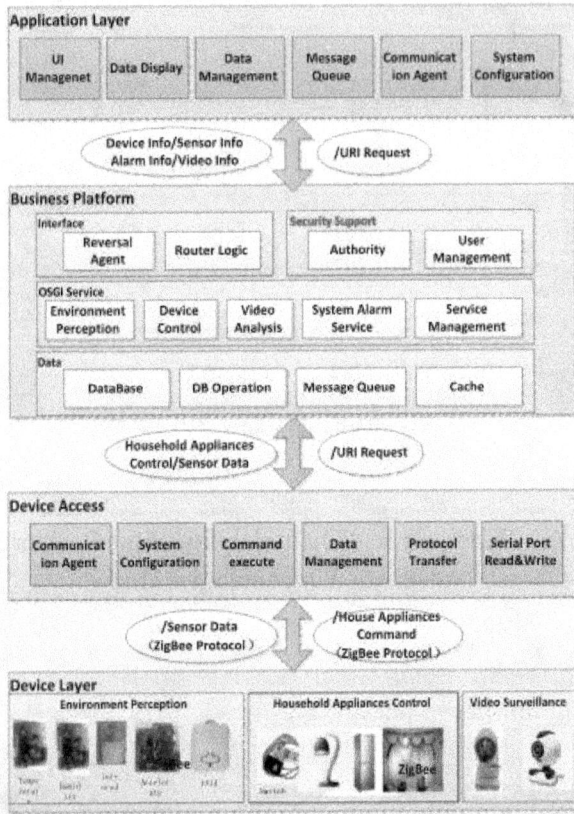

Figure (6). Smart home system architecture [10]

Every sensor will be attached to one of the appliances. Thus, these sensors can connect all the devices with the home gateway to upload the data or to accept a new command from the stakeholder.

The gateway here plays an essential role by connecting all the sensors to the business platform, collecting data, distributing new commands, and converting the context of messages when dealing with different communication protocols. The service provider is the business platform in the system. The main purpose of the business platform is to manage the data as well as to provide services and logic processing. In addition, it provides a unified URL interface to be displayed on the user's mobile devices as a smart terminal, or on the gateway side. There are three benefits from the user's terminal, which include dynamic interfacing, retrieving data, and applying new commands to the business platform using smart terminals. From Figure 3, there are different varieties of communication protocols used by terminals that hold the retrieved information coming from the home gateway. However, the business platform overcomes the heterogeneity of smart terminals by providing a unified access interface and understanding the undifferentiated procedure with a smart terminal and home gateway [10].

SMART HOME SERVICES

HOME APPLICATION AREAS

In the last decades, smart homes have been a very interesting field for researchers. Many of the studies were on how to monitor and control different zones in one home. The classification goes to four different areas of interest, which include Security, Energy Efficiency, Heath Care, and Entertainment [13]. According to the management of energy efficiency and power control, a smart home is located in the middle of a smart grid and the consuming devices, so creating a smart gateway to manage the power consumption in a home completes the entire goal of a smart gird network. In this way, the government's policy-making employers can create the benefits of optimizing the consumption by users while reducing the cost of an electrical unit. Therefore, many solutions were found to encourage the use of smart metering, which has the ability to communicate with both the smart gateway and smart grid. In other words, there are two ways to calculate home power consumption. The first is to use a regular meter; the second one is to monitor every device using a smart meter, which means using distributed devices in the entire home and giving the information to the main unit. Some people see that the second type is complex to implement and use since it will be very costly to wire. However, it will be easy to overcome these difficulties by using wireless technology [17].

PREDICTION VIA SMART SYSTEM

In general, modern technology, which develops very fast, causes people to try to delegate many of their needs to smart machines. These machines are built with the anticipating capabilities to predict what the users intend to do in a certain environment like a smart home. In [18], the authors indicate that "The more complex the operating system is, the more it is expected to do on behalf of its users." For that reason, many researchers use artificial intelligence, which involves many techniques. These techniques may include analyzing the user's actions from the database then doing some stochastic operations to try to predict the next user action. Therefore, the smart homes must be an adaptable and dynamic intelligent system, which means it needs to adapt to the stakeholder's lifestyle and to anticipate his or her next activities to optimize the direct interface between the user and the home appliances. Knowing the lifestyle of a particular user in a home requires precise knowledge about the history of appliance usage for each state in the entire environment. There are many methods to anticipate the user's future actions, such as intelligent Human Computer Interaction (HMI). The principle of the prediction process is studying the sequence of interaction events to generate the next predicted user action or state [19].

The following paragraphs discuss a case study for a new approach of a smart home as a basic element of smart cities that have been supported by using the Internet of Things' design pattern.

CASE STUDY: NEW DESIGN APPROACH OF SMART HOME

The Functional Organization of the smart home system, as shown in Figure 7, introduces four main categories of applications that have been supported through the building of a smart environment in our homes. Figure 6 shows how each sensor, actuator, and device (nodes) in the designed system can be classified according to its functionality. For instance, we can see that the security box has many security nodes, which have a similarity in their functionality. Examples are entry door sensors, motion sensors, smoke sensors, and CCTV cameras. The healthcare box has the nodes that are concerned about personal health for all people but especially for elderly or disabled people. All these boxes are connected to a gateway. Many researchers present the functionality of the smart home gateway as an interface component to collect, monitor, and notify the stakeholder about the state via specific communication techniques as presented in [12], [13], [15], [20]. Further development has been proposed not only to monitor the status but also to implement the received order that been initiated by the stakeholder himself for a current status via a gateway as mentioned in [21], [22]. However, the smart gateway that has been presented by many researchers should be designed to do a lot of work on behalf of the user. These activities may include but are not limited to keeping track of devices, analyzing the data, and visualizing it in the form that makes it easy for the user or a proficient partner in the application layer to monitor the status then personally control the state of any nodes as needed to the preferred values.

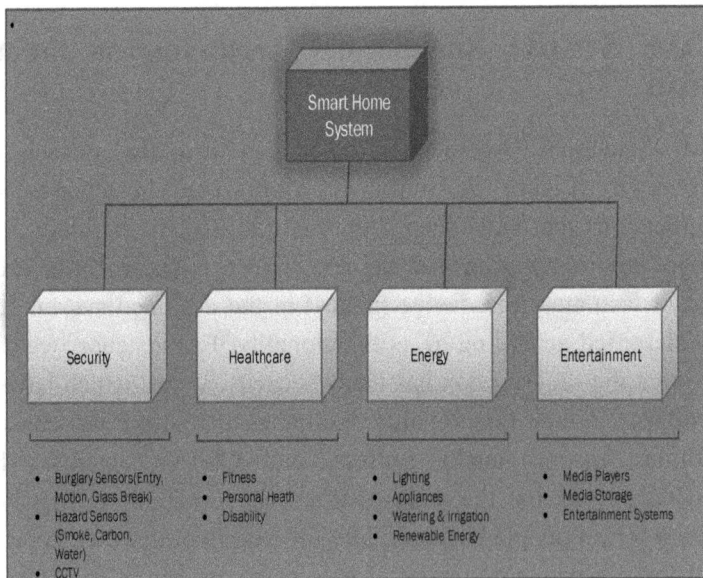

Figure (7). The Functional Organization of the Smart Home System

Nowadays, the IoT gets a lot of attention from researchers and very big companies like Google, Amazon, Microsoft, Xively, and many more. The IoT plays an important role in monitoring anything, anytime, and anywhere. As can be seen in Figure 7, the smart home based on the IoT is an approach that uses the IoT platform to get services. Thus, the stakeholder in the application layer needs to use business services, device access, and application partners that enable businesses to quickly connect products and operations to the Internet.

The researcher in [10] presents a smart home as "A typical application scene of Internet of Thing, recently has attracted wide attention from industry to academic circles." However, they present many issues due to using many heterogeneous devices in the home environment. These issues are related to integrity, universal data management, and how to work with each other in case of different kinds of subsystems.

In [23], the researchers mention that the complexity and the difficulty that we can find in the new emerging technology creates large challenges for further adaptation for centralized network management. This is because of the heterogeneous behavior of such a network as well as the use of a variety of services, objects, and applications from more than one vendor. Therefore, this kind of network is assumed to be neither efficient nor scalable anymore. The author also adds that in "today's real networks, there are many network management complexities and limitations that cannot be adequately solved by a fully centralized approach, such as lack of flexibility and information bottlenecks." So, the complexity and distributed electronic environment in our home needs a new approach or a distributed approach, which means that the designed network is logically centralized but can be physically distributed.

The proposed system, as shown in Figure 8, tries to enhance the design of the smart home environment by introducing a novel approach represented by using a smart agent for each subsystem, a storage agent with the cloud for backup data with an enterprise analysis, and a central intelligent agent (Brain) that connects all the components of our proposed system in a one unique design approach, which we call the BUTLER. The idea behind this design is to present a novel approach of a personal assistant in the home. As shown in Figure 9, the intelligent agent represents the brain of the home assistant while the subsystems and their related agents represent the human senses.

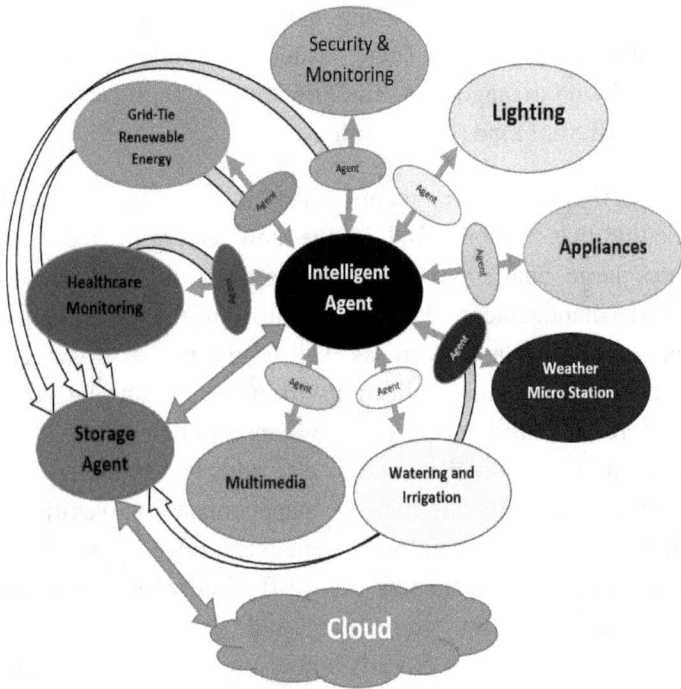

Figure (8). The Intelligent Home System Design

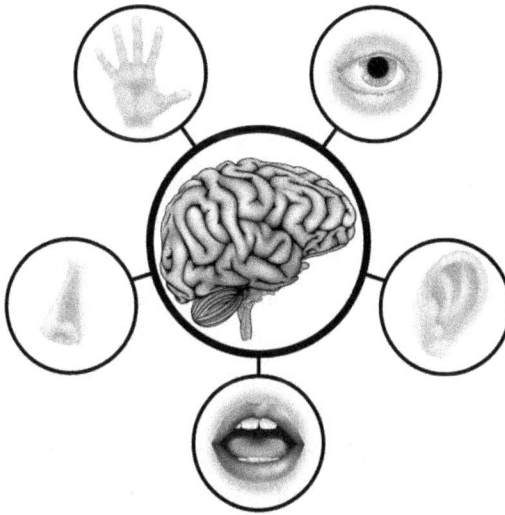

Figure (9). The Five Sense Organs in Human Beings [24]

Our approach presents a new idea or approach for a smart-house system that can learn the user's behaviors and then independently perform tasks for the user, often anticipating the user's needs. In the same time, the design presents an integrated paradigm to the green architecture design approach, in addition to presenting an integrated platform for the Internet of Things paradigm to handle wire/wireless sensing systems and information processing efficiently.

In the following paragraphs, we discuss the necessary steps to present our intelligent system. These steps include the requirements analysis of the system, system design, the architecture design, finally the implementation of the system via NetLogo simulation program.

System Requirements

The system should have the ability to interface with any electrical or electronic device in the home. It must be capable of communicating using a variety of different standard communication protocols to provide interoperability. It is easy to maintain. In addition, it also has the ability to collect the data from the environment at any time then analyze this data to make a decisions depending on similar hypotheses from previous rules. In other words, the system has the ability to self-learn, is autonomous and adaptable to any change in user behavior in any time at any place in the home. So the system tracks the behavior for each user, stores and retrieves the behavior to/from the short term (Local) memory and long term (remote) memory in the cloud. Finally, the system should provide different types of user interface such as a touch screen unit, microphone for voice command, internal speaker and should have the ability to send/receive SMS or MMS to provide more reliability to the designed system. Referring to Figure 7, the design shows all the necessary components of the system, which includes subsystems, agents, and Intelligent Agent (IA). Each component has its own independent requirements.

System Design

To manipulate all these requirements, the design of the system is proposed to consist of eight subsystems in addition to the central Intelligent Agent Unit shown in Figure 7. Each subsystem has different kinds of sensors, actuators, and devices which we call NODES. For each subsystem, all the nodes should be connected to the subunit of that subsystem. For example, the Multimedia Subunit connects all the multimedia devices in the home. In addition, every subsystem has an independent agent. This agent plays an important role in communication between the subunit and the central Agent (IA). Every agent can communicate with its subsystem using the same message language protocol that this subsystem uses. The language protocol between the IA and all the agents are the same, whether the communication protocol is wired or wireless. The central IA connects all the subsystems as one network, which is shown in Figure 7. This IA, as we compared to the Five Sense Organs in Human Beings in Figure 8, represents the brain of the automated system that continuously works to track the user's behavior by saving the information in the database, minimizing it, then analyzing the data to get the information. After producing some information about the preferences of the user, the next important point is to predict the user's needs and act on behalf of the user himself/herself to do the delegated action.

Now, we examine how the specific design that is discussed above meets the system overall requirement, Figure 10 shows the architecture design of the two suggested subsystems that communicate with the IA, which control and manage the message flow for all the messages coming from or going to any subsystem.

From Figure 10, we can clearly see that each subsystem has its own specific nodes that differ from other subsystems; hence, the security subsystem has many types of nodes such as motion, smoke, door lock, window break, or CCTV, while the lighting subsystem has different kinds of nodes such as illuminance sensors, lighting bulbs, and LED. All these nodes are connected to a security subunit. The subunit connects all security sensors, actuators, and devices to a security subsystem agent using either wired or wireless standard communication protocol. It is important to mention that there are many tasks assigned to each agent, which also differ from one agent to another. These tasks may include monitoring and controlling all security nodes, buffering, reading statuses to send them to the intelligent agent, running predefined commands from the IA, and doing some data mining and context aware processing.

A similar task in the lighting subsystem agent is implemented to deal with different kinds of lighting nodes. The subsystem's agents, as seen in Figure 9 take the block numbers 2 and 4. These agents are located between the subsystem and the IA. The agents here try to read the nodes' status from the subsystem's subunit by pulling an operation process and saving the data in its local memory. After that, it analyzes the data to get information then sends the information that was extracted from the redundant data to the IA. In some cases, the redundant data with its information that the agent extracted is sent directly to storage agent block number 6 to be saved in the cloud block number 7 as backup data for the system and for further analysis. When any change happens in any node's status, it should be sent directly via the agent to the brain of the entire system, the IA to take certain   action.

At the same time, the agent who issues the notification message to the IA should continuously read and implement the incoming new command via the pulling process for its local memory. It is important to mention that the agent plays an important role between two different communication protocols. Since the communication protocols between the agent and the subsystem may differ from the communication protocol that has been used to connect the agent with the IA, the agent here takes care of the context aware message protocol. This point assures the compatibility and open architecture design between different communication systems; as a result, this point makes the system's   extendibility   much   easier   to   perform.

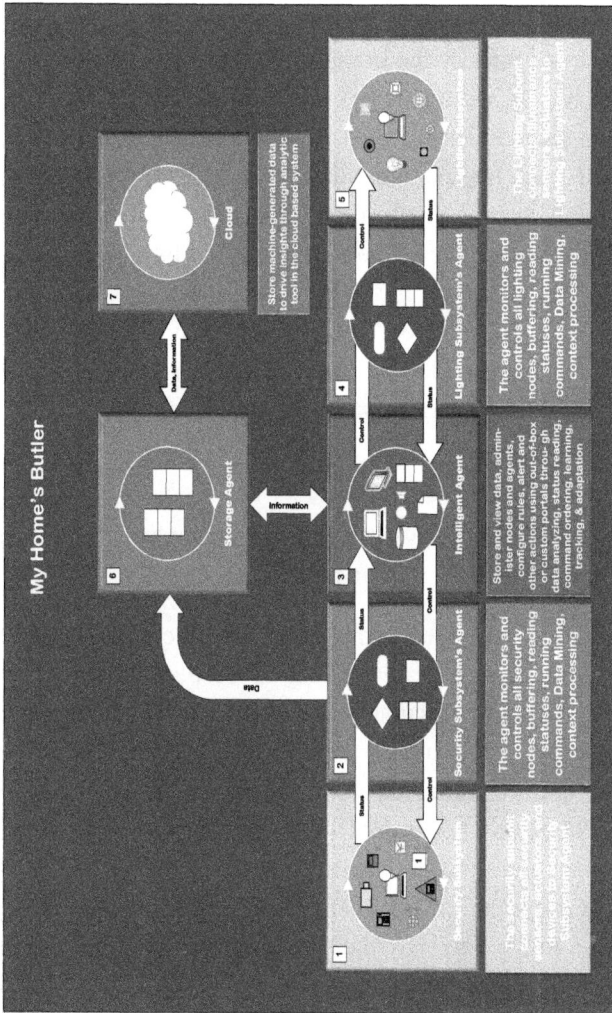

Figure (10). The communication and information flow between the IA, two subsystems and cloud.

The IA here plays the most important role between the butler's components. It is basically the brain of the designed system. This brain has a lot of work to do, such as connecting all the subsystems' agents and the storage agent with the cloud to the central unit. It must be reactive, autonomous to take the control over the other subsystems through self-learning, flexible, and adaptable to user behaviors. The storage agent here proposes a firewall agent. It can be supported with firewall capabilities as a security supported agency or isolation gateway to perform a specific task, such as isolating the inner environment (home environment) from the outer environment (cloud system).

COMPARISON BETWEEN MODE OF OPERATION (MANUAL MODE AND AUTOMATIC MODE)

The manual mode here means the direct interface between the user and the node itself or what we do in our daily life; in another words, using the regular nodes and electrical outlets that we use every day that have nothing to do with the smart environment, like a tungsten lamp plugged into a simple electrical outlet or any appliances like a regular HVAC system. The Automatic mode represents the mode where the butler is being used in the same environment to do some jobs that have been delegated to the butler by the stakeholder. Doing the job in an efficient way demands using some types of artificial intelligence. To reach this level of intelligence, this mode has two previous steps that must be implemented before reaching the current level. All these steps are discussed in the following section.

SYSTEM DEVELOPMENT PHASES FOR AUTOMATIC MODE OF OPERATION

The automatic mode needs a much more complex system than the simple system in the manual mode. This system is divided into software and hardware components. The hardware component includes the agent for each subsystem, the storage agent, the IA, and the cloud based system. The software component of this system should support both standard wired and wireless communication protocols, the necessary preconfigured software programs for each agent, in addition to the operating system for the intelligent system.

The automatic mode of operation, as shown in Figure 11, has three phases: Initialization Phase, Learning Phase, and Action Phase, as discussed in the following points.

INITIALIZATION PHASE

This phase includes all the initialization parameters for the nodes and the related subsystem agent software programs for each agent, in addition to the intelligent agent itself. The initial parameters and software programs are necessary when any part of the system needs to be restarted or rebooted. Many of the parameters and rules that are presented as initial rules for operation from the starting point are actually present in the default rules that any system begins with. These rules are basically the preliminary procedures that the intelligence is going to use when certain events occur at any time, such as alarm detection or rainy weather. All these configurations are necessary to begin the next step or what we called the learning phase.

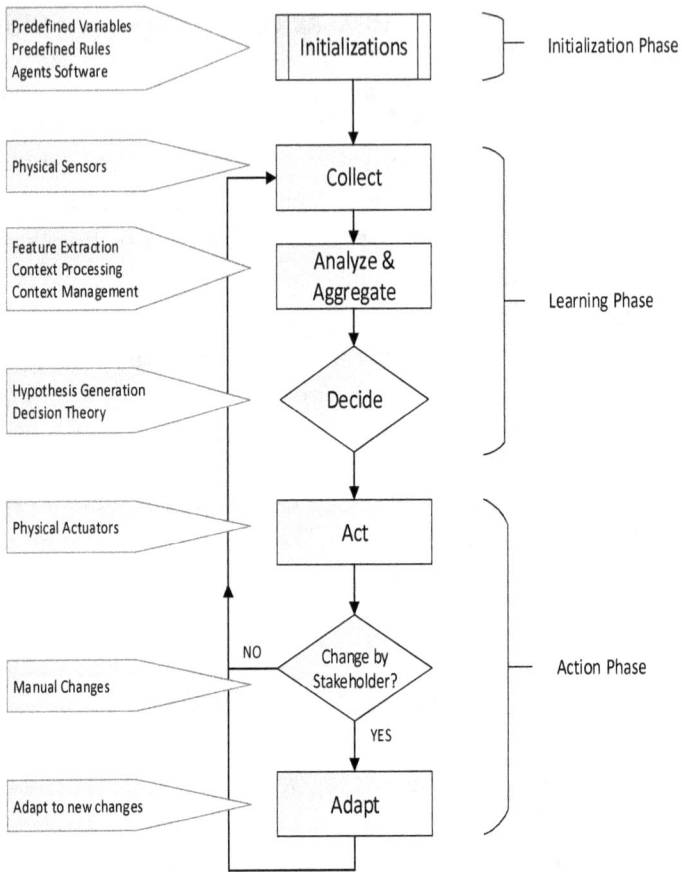

Figure (11). The development phases for the automatic mode
during the active mode of operation

LEARNING PHASE

This represents the second phase of the automatic mode of operation. It consists of four essential steps that begin with collecting the status for each physical node from all the subsystem agents in the IA local memory. The next step is to analyze and aggregate the information that comes from the agents using some feature extraction, context processing, and context management. The decision step presents the use of decision theory that is used to generate the hypothesis for current inputs. The hypothesis here represents the target function that the IA wants to model for the input information. If this hypothesis is new for the IA, then the IA should wait for the user to take an action. After this step, the IA can build or generate a new rule to use it in the future for the same hypothesis. If the current hypothesis is similar to a hypothesis that exists in the local memory, then the IA should act depending on the user demands. This phase is very important because it tries to build a new rule that consists of several input variables coming from node statuses and the output, which means the stakeholder's action according to that input. The strategy to build new rules comes from tracking the user's behaviors against different node statuses in the surrounding environment and from implementing these rules in the last phase of the automatic mode. The new rule is used along with the initial rules to take control of the system on behalf of the user in the next phase.

ACTION PHASE

In this phase, the IA tries to take the right action using the applicable rule for the current state or node status. This action is represented by the Act block in Figure 10. The last block is labeled Adapt, which means that the IA should have the capability to update the learned rules from the previous phase due to an unexpected change that has been done by the stakeholder manually. In another words, the adaptation here means the changing or editing process for the produced rules in the previous phase that the IA should follow.

These changes are actually due to the changes in user behavior (the output variable for the target function) for the same input node statuses (the target function's input variables). It is important to mention that this adaptation happens after the IA detects the same user action three times concurrently. In other words, if the change in stakeholder behavior happens just one time, this change is considered in the related rule.

This phase is very necessary to implement, since the human, for some reason, may change his mind for a short period of time, such as housing a guest or responding to changes in weather. Continuously, the IA follows these changes and tries to edit the needed rules to satisfy the stakeholder's needs and to perform the delegated goals that have been given to the Butler.

CONCLUSION

A smart home's design, as discussed in previous sections, definitely helps people who are interested in this field of knowledge through presenting and analyzing important points for a robust design such as:

- Effectively integrates heterogeneous communication systems and information processing systems for measuring the normal condition of people living independently.
- Provides the usage behavior of different types of domestic nodes inside the home for different users.
- Customizes the behavior of nodes to every independent usage based on the users' requirements.
- Shows the collaboration and weights between multiple subsystems that have different functionalities on one integrated system called the Butler.
- Builds a home's knowledge by drawing a relationship between heterogeneous smart nodes for individual daily behaviors.
- Presents a human behavior model via analyzing the real sequence of individual activities and probable usage of household appliances.
- Provides interoperability between different types of components using an agent based system.

REFERENCES

[1] R. Y. Clarke, "Smart cities and the internet of everything: The foundation for delivering next-generation citizen services," Alexandria, VA, Tech. Rep, 2013.

[2] I. Zaman, "white paper on smart cities," 2015.

[3] A. Z. Angelo Cenedese and L. Vangelista, "Padova smart city: an urban internet of things experimentation," World of Wireless, Mobile and Multimedia Networks.

[4] I. Gartner, "Urban efficiency as the cornerstone of attractive cities," Schneider Electric.

[5] e. a. Harrison, C., "Foundations for smarter cities. ibm journal of research and development," IBM Journal of Research and Development.

[6] L. Arrowsmith, "Smart cities: Business models, technologies and existing projects world 2014," IHS Technology.

[7] A. Al-Ali, I. Zualkernan, and F. Aloul, "A mobile gprs-sensors array for air pollution monitoring," Sensors Journal, IEEE, vol. 10, no. 10, pp. 1666–1671, 2010.

[8] X. Li, W. Shu, M. Li, H.-Y. Huang, P.-E. Luo, and M.-Y. Wu, "Performance evaluation of vehicle-based mobile sensor networks for traffic monitoring," Vehicular Technology, IEEE Transactions on, vol. 58, no. 4, pp. 1647–1653, 2009.

[9] A. Zanella, N. Bui, A. Castellani, L. Vangelista, and M. Zorzi, "Internet of things for smart cities," Internet of Things Journal, IEEE, vol. 1, no. 1, pp. 22–32, 2014.

[10] L. Sun, H. Ma, D. Fang, J. Niu, and W. Wang, Advances in Wireless Sensor Networks: The 8th China Conference, CWSN 2014, Xi'an, China, October 31–November 2, 2014. Revised Selected Papers. Springer, vol. 501, 2015.

[11] N. K. Suryadevara and S. C. Mukhopadhyay, Smart Homes. Springer, 2015.

[12] L. C. De Silva, C. Morikawa, and I. M. Petra, "State of the art of smart homes," Engineering Applications of Artificial Intelligence, vol. 25, no. 7, pp. 1313–1321, 2012.

[13] T. Mendes, R. Godina, E. Rodrigues, J. Matias, and J. Catal~ao, Smart Home Communication Technologies and Applications: Wireless Protocol Assessment for Home Area Network Resources, vol. 8, no. 7, 2015.

[14] E. Borgia, "The Internet of Things vision : Key features , applications and open issues," Computer Communications,

vol. 54, pp. 1–31, 2014.

[15]N. K. Suryadevara and S. C. Mukhopadhyay, Smart Homes: Design, Implementation and Issues, 2015.

[16]C. Withanage and K. Otto, "A Comparison of the Popular Home Automation Technologies," IEEE Innovative Smart Grid Technologies - Asia (ISGT ASIA) A, pp. 600–605, 2014.

[17]F. Viani, F. Robol, A. Polo, P. Rocca, G. Oliveri, and A. Massa, "Wireless architectures for heterogeneous sensing in smart home applications: Concepts and real implementation," Proceedings of the IEEE, vol. 101, no. 11, pp. 2381–2396, 2013.

[18]P. B. Galvin, G. Gagne, and A. Silberschatz, Operating system concepts. John Wiley & Sons, Inc., 2013.

[19]M. Marufuzzaman and M. B. Ibne Reaz, "Hardware simulation of pattern matching and reinforcement learning to predict the user next action of smart home device usage," World Applied Sciences Journal, vol. 22, no. 9, pp. 1302–1309, 2013.

[20]J. Bian, D. Fan, and J. Zhang, "The new intelligent home control system based on the dynamic and intelligent gateway," in Broadband Network and Multimedia Technology (IC-BNMT), 2011 4th IEEE International Conference on. IEEE, pp. 526–530, 2011.

[21]B. Lee, S. Yang, and D. Choi, "A status monitoring system design/implementation for home appliances controlled by home server," 1st International Conference on Networks and Communications, NetCoM 2009, pp. 220–223, 2009.

[22]M. Wang, G. Zhang, C. Zhang, J. Zhang, and C. Li, "An IoT-based Appliance Control System for Smart Homes," pp. 744–747, 2013.

[23]J. Szyma, Z. Chaczko, and B. Roda, "Advanced Methods and Applications in Computational Intelligence," vol. 6, pp. 387–404, 2014.

[24]http://project4cre8tiveintent.com/2015/09/22/five-d-harder-65

CHAPTER EIGHT

SOCIAL SENSOR NETWORKS

Basim Mahmood[1] and Abduljaleel Al-Rubaye[2]

[1] Department of Computer Science, University of Mosul/ Iraq

[2] Department of Computer Sciecnce, Florida Institute of Technology, Melbourne, FL, USA

ABSTRACT

This chapter presents and describes the integration of Wireless Sensor Networks (WSNs) and Social Networks. Nowadays, WSNs have caused a paradigm shift in our society. They have become a popular means of communication among people. Many aspects of our lives are significantly related to the WSNs such as communication, transportation, military, and agriculture. The main concern when designing WSNs is the limitation in network resources (e.g., memory and battery) at sensor nodes. We live in a society where almost everyone carries a sensor in the form of a smartphone or a tablet. This ubiquity means that mobile WSNs already exist around us. Furthermore, the mobility of people may be able to be exploited in the design of more efficient WSNs because the sensors would not consume energy to achieve movement.

Another interesting aspect that emerges from the scenario above relates to social interactions formed from the encounters of sensors: Social Networks. In Social Networks, the relations among nodes represent the main source of structural information about the phenomena they represent. The collective values of a network relations can further help us to understand how information flow within a network. This chapter focuses on challenges and issues that relate to the intergration of WSNs and Social Networks (Social Sensor Networks), such issues are: memory requirements and memory management.

Keywords: Wireless Sensor Networks, Social Networks, Mobility Models, Smart Cities, Internet of Things (IoT)

INTRODUCTION

A Wireless Sensor Network (WSN) is a collection of sensors that are mainly used to measure environmental or physical conditions. WSNs have been widely used in many applications such as forest monitoring, infrastructure protection, and mobile object tracking. The rapid growth in the use of WSNs has yielded many challenging tasks such as efficient memory usage, power consumption, security, and connectivity. The structure of a WSN can be static or dynamic. In static WSNs, each sensor is in a stationary position; the performance depends on factors such as location or communication range. In dynamic (mobile) WSNs, sensors' positions are subject to change over time. This change in position, however, affects the consumption of network resources (e.g., memory or battery). In these networks, a Mobility Model can characterize the movement's behaviour in terms of direction and speed for each sensor. This model can be incorporated into a dynamic WSN for simulation and evaluation purposes.

The use of mobility models in WSNs yields expectations for the costs relating to sensors' behaviour in movement. Hence, the study of mobility behaviour can help us in determining which mobility model is ideal for a particular application. Most mobility models used with WSNs are based on random movements, such as the Random Waypoint Model and the Levy-flight Model. However, other mobility models can be coupled with WSNs, for instance, models of human mobility are considered non-random and can be coupled with WSNs. The coupling of sensor nodes to humans may be interesting due to the fact that nodes would move as a consequence of human movement; no mobility cost would be incurred to the WSN. When attaching sensor nodes to humans, an issue that becomes apparent is the effect of individuals' social network to the connectivity of the WSN.

Another issue that may be important to real-world WSNs relates to the structure of human connectivity. Social networks describe the social relations among network actors (e.g., individuals or groups). A social network can be represented as a graph where its vertices represent actors and the edges represent social relations among actors [1]. The study of social networks can help in understanding the structure of social relations among people and their effect on social movement patterns (mobility). Therefore, the analysis of social networks is useful insofar as human mobility is likely to be influenced by people's social ties.

Social network structures can reveal the ties (relationships) among individuals, which in turn can help the understanding of how information flows within a network [2]. Basically, there are two main types of ties [3]: Strong Ties and Weak Ties. Strong ties exist among family members, friends, and people with whom we associate frequently, while weak ties exist with whom we associate infrequently.

In this context, the concept of Social Capital has been introduced as an indicator of the collective value of a particular individual's tie within a social network [4]. The quantification of the concept of social capital is a useful tool when it comes to the general importance of ties in a network.

As we can see from the description above, the concepts of human mobility and social interactions are intertwined. In today's world, we live in a society where it is more likely that everyone carries a sensor in the form of a smartphone or a tablet. These sensors move as a consequence of human movements. This ubiquity means that mobile sensor networks already exist around us; and thus sensor mobility strictly reflects human dynamics behaviour (i.e., sensors move with the human carrier). Another interesting aspect that emerges from the scenario above relates to social interactions formed from the encounters of sensors: social networks.

The integration of the aforementioned fields forms a social framework in which people encounter each other as a part of their social activities. Tracking these encounters requires devices to memorize their history of encounters. Therefore, each device in this social framework creates a history of encounters with the other devices in its memory in order to further use this information for different purposes such as memory management and information dissemination. Figure 1 depicts the general structure of the social framework we are dealing with, which can be called Social Sensor Network.

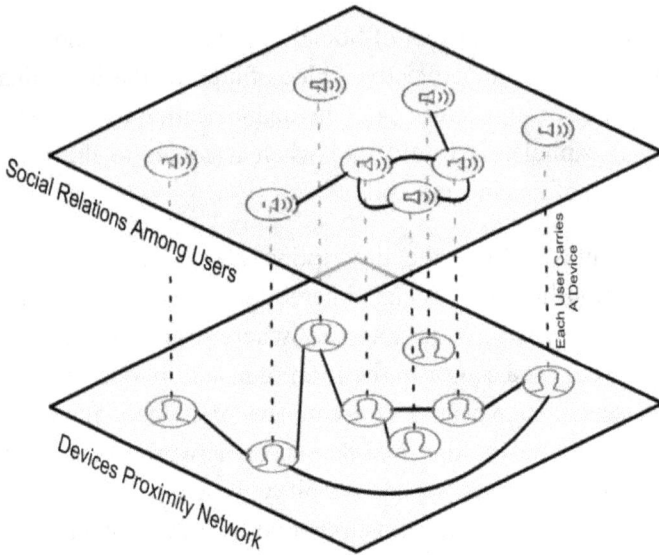

Figure (1). The general structure of Social Sensor Network. There are two levels of connectivity: social relations and proximity. Sensors can send messages to each other when they are in close proximity. The top layer in the figure is more dynamic because users are mobile. The social relation layer is also mutable but the social relations do not tend to change as often as in the proximity network.

WIRELESS SENSOR NETWORKS: CONCEPT AND STRUCTURE

WSNs have become an important area of research [5,6]. Although this area is a relatively new field of research, it significantly relates to many different fields and applications such as communication, transportation, military, and agriculture [7]. A typical WSN consists of many small devices distributed over a geographical area, where each device is called a Sensor and can measure environmental or physical conditions in a particular area of interest.

In a WSN, each sensor node can be characterized as follows: node size, varying according to their purpose; node cost, differing from one sensor to another depending upon their complexity; communication range, varying from a few centimetres to thousands of meters depending upon the sensors' hardware features; storage capacity, limited and restricted by its hardware size, ranging from a few kilobytes to several megabytes or even gigabytes (e.g., a cellphone). Similar to other communication devices (e.g., network interface cards), each sensor in a WSN has a unique identifier that is assigned by the manufacturer. This identifier is so-called Media Access Control Address or MAC address. This address is used to distinguish a particular sensor among many. Assigning a MAC address to a particular device follows the standard IEEE 802 format, consisting of 48-bits (6 bytes).

In a WSN, each sensor has to be able to perform two main functions. First, it must be able to observe events in its communication range. A better network performance can be gained when sensor communication range is relatively high.

Second, it must be able to connect to other sensors in the network, which is important for information flowing within the network. Yet, sensory information should be transferable to other sensors [8].

Typicaly WSNs have two main classes of sensors [9]. Sink nodes work as base stations that collect sensory information from other nodes; this class of node has high capabilities in terms of energy resources, processing speed, and memory capacity. Regular nodes are the other class of node; they are used to observe events within an area such as temperature. Structurally, the way of distributing sensors (topology) over an area is based on the network purpose.

Since a WSN consists of a number of nodes with connections among them, it is possible to use the graph theory to represent the physical structure of a WSN. According to the graph theory, a graph representing a WSN is given by $G = (V, E)$, where V denotes sensor nodes, and E denotes connections among nodes.

ROUTING IN WSNS

Routing is the process of choosing the most ideal paths for sending data (e.g., a message) from its source to the final destination [10]. This optimization process, however, is considered as one of the challenging tasks when designing routing protocols [11]. The main idea behind routing in WSNs is to reduce the consumption of network resources (e.g., power and memory) when transferring data from one node to another [11]. There are many classes of routing protocols as follows:

FLAT-BASED ROUTING:

This class is used in networks with a large number of sensors. Sensor nodes are given equal roles and collaborate to perform a particular task [12]. This class of routing prevents delivering redundant data within a network.

HIERARCHICAL-BASED (CLUSTERED-BASED) ROUTING:

Network nodes perform different functions. Transferring and processing tasks are assigned to high-capability nodes (e.g., sink nodes), while regular nodes can be only used for sensing tasks [13].

LOCATION-BASED ROUTING:

Uses nodes' location to limit the area for finding a new route for sensory information. If there is no activity in a sensor's area, then sensors in that area should go into a sleep mode [2].

Furthermore, there are many routing protocols have been proposed such as Epedimic routing protocol [14], PRoPHET routing protocol [5], Spray-and-Wait [16], Gradient-Based routing protocol [17] to mention a few.

SOCIAL NETWORKS: CONCEPT AND STRUCTURE

A social network represents the structure of relations among a set of actors such as individuals, organizations, or groups. The study of social networks is a multidisciplinary discipline composed of core concepts that can be used in a multitude of disciplines including biology, economics, and geography to mention a few.

The analysis of social networks [1] combines statistics, sociology, social psychology and graph theory. The study of social networks can help us to understand the relations among people and the behaviour of actors in a network.

Social networks can be represented as a graph [18]. A social network is a graph G = (V, E), where V denotes network nodes, and E denotes the relations among nodes. The relations among nodes (actors) in a social network can be directed or undirected. For a better understanding, consider a set of social actors {1, 2, ..., n}, and (i, j) representing that actor i is linked to actor j. If we assume a directed relation between both actors then (i, j) ≠ (j, i). In an undirected relation, the relation (i, j) = (j, i). Furthermore, a relation between two nodes (i, j) can be weighted based upon the strength of the relation between i and j. Figure 2 shows the graph representation of a network.

Each node in a network can be characterized by many features such as a node degree, which is the total number of ties to other nodes. In directed networks, a node has an in-degree that representing the number of incoming ties, and an out-degree that representing the number of outgoing ties. The degree distribution of a network is the probability distribution of degrees over the entire network [19] where it is an important aspect in analysing real-world networks (e.g., the World Wide Web and Online Social Networks). In these networks, some nodes are highly connected (hubs), but other nodes have few connections.

Therefore, the degree distribution in such cases follows a power-law distribution. This type of network is known as Scale-Free-Networks [20].

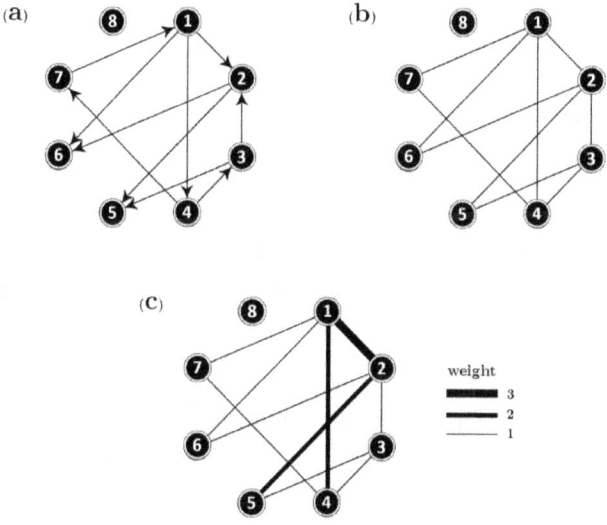

Figure (2). Three types of networks. Each one of the networks consists of 8 nodes {1,2,3,4,5,6,7,8} and 10 edges. Network (a) shows a directed graph, where communications flow in one direction. For instance the edge (1,2) is not the same as (2,1). Network (b) is undirected; all the links run in both directions. Network (c) represents a weighted graph, where the edges (1,2), (1,4), and (2,5) have more dominant weights in comparison to other edges' weights.

In social networks, nodes can be clustered into groups or communities, which share common features based upon common characteristics, such as position and function [21]. For example, citation networks [22] contain communities by research topic; protein-protein interaction networks [23] can have communities of protein with the same functions. In these cases, a network is said to have a community structure.

In addition to the aforementioned characteristics, social networks relations also have other characteristics such as:

TIE STRENGTH:

Ties in a social network can be Strong or Weak. Strong Ties can be formed with people we associate with frequently (e.g., close friends or family members), while weak ties with the people we associate with infrequently (e.g., people we meet at a resturant).

HOMOPHILY OR ASSORTATIVITY:

The tendency of nodes to associate and connect with similar others [24,25,26]. According to a study by McPherson et al. [26], there are two main aspects of homophily, namely; status homophily, referring to the fact that individuals with similar social status features are more likely to participate with each other by chance (e.g., race, gender, religion, or age), and value homophily, meaning individuals tend to participate with those who behave and think in similar ways regardless of differences in social status. However, value homophily has a more significant impact on homophily than status homophily as presented in the study of Yuan [27]. Moreover, many factors cause homophily such as geography (e.g., people locations), social ties (e.g., strong and weak ties), organizational foci (e.g., school or work), isomorphic sources (e.g., people who occupy equivalent roles), and cognitive processes (e.g., people who have demographic similarity).

TRANSITIVITY:

When there is a relation from node i to node j, and from node j to node k, then there is also a relation from i to k (e.g., friends of my friends are also my friends) [28].

RECIPROCITY OR MUTUALITY:

The tendency of actors to form reciprocal ties (double links) among each other (e.g., among friends) [29]. In order to know how important a node is in a social network, one can use centrality measures. The centrality measures of a network can be applied at two different levels: network level and node level. There are many measures of centrality such as:

Degree Centrality, representing the total number of ties to other nodes in a network. This measurement can only reveal how many connections a node has [30].

Betweenness Centrality, representing the total number of shortest paths that pass through a particular node in a network. This measurement can show nodes that are more likely to receive information in a network [31].

Closeness Centrality, representing the sum of distances to all nodes in a network. A node has a high closeness in a network if it is relatively close to all other nodes [30].

MOBILITY MODELS

In dynamic networks, the term Mobility represents the description of how network nodes behave in terms of direction and speed. More precisely, a mobility model in a dynamic network describes the movement of mobile nodes and how their positions, directions, and speed change over time [32,33]. Typically, to simulate and evaluate a dynamic WSN, a particular mobility model should be incorporated into the network [34]. Brownian motion (BM) [35], which was observed for the first time by Robert Brown in 1827, then later discussed in a more precise detail by Albert Einstein in 1905, was the first mobility model investigated by scientists. The BM describes the motion of particles suspended in a fluid (e.g., gas) as a result of their collision with the atoms in a fluid. Einstein showed that the probability for a particle at a distance r from its initial location after time t follows a normal distribution.

The research community in this field has contributed in proposing several mobility models with different characteristics such as the following models:

LEVY-FLIGHT:

A random walk model in which the step length among jumps is not constant and follows a heavy-tailed distribution [36]. The probability P(Δr) of finding a step of size Δr is:

$$P(\Delta r) \simeq |\Delta r| - 1 - \beta, \qquad (1)$$

where $0 < \beta < 2$. Therefore, Levy walks consist of many short jumps and occasionally long jumps. It can be observed in animal foraging behaviours (e.g., the behaviour of foraging ants). According to the distribution of step lengths among jumps, this random walk may be called Cauchy flight if the step lengths follow a Cauchy distribution and Rayleigh flight if they follow a normal distribution. Figure 3 shows a comparison between the trajectories of two types of random walks, Levy and Normal.

RANDOM WAYPOINT MODEL (RWP):

Proposed in 1996 by Johnson et al. [37], simplicity is this model's feature. As this model starts, each node selects a location randomly as its destination in the environment. It then moves towards the destination with the speed selected randomly and uniformly from [0, Vmax], where Vmax denotes the maximum permissible speed for every node in the network. The direction and the speed of the node are selected independently. As soon as the node reaches its destination, it stops for a period of time, which is determined by its waiting time Twait. Thereafter, the node randomly selects a new destination and repeats the same process. Clearly, the mobility behavior of this model is mainly controlled by the parameters Vmax and Twait. In this context, if Twait is small and Vmax is large, then the network that uses this mobility model is considered to be highly dynamic, whereas, when Twait is large and Vmax is small, the network is expected to be stable [38]. Therefore, different scenarios can be generated by varying these parameters. However, this model does not reflect all real life scenarios. For example, the walking speed of humans is incrementally increased and the direction is smoothly changed.

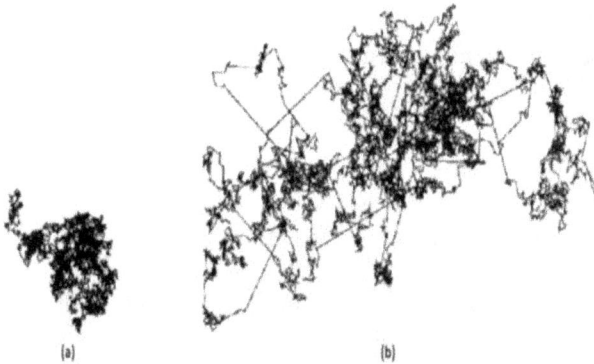

(a)   (b)

Figure (3). Comparing the trajectories between (a) in which the step lengths. Follow a normal distribution and (b) Levy walk. Both walks have ≈ 7000 steps with $\beta = 1.5$ [39].

RANDOM WALK MODEL:

Proposed by Pearson [40], in this model, at each time interval, mobile nodes modify their speed and direction. At every new interval t, each mobile node selects its new direction $\theta(t)$ uniformly and randomly from $(0,2\pi]$. The new speed $v(t)$ of each node is selected from $[0,Vmax]$. In this model, the current speed of a node is completely independent of its previous and subsequent speed. Basically, this model is close to the RWP model with one difference: there is no waiting time among jumps [41]. Each node can freely move with no restrictions to any location. However, in urban areas, people's movements are restricted by many barriers or obstacles (e.g., buildings). Therefore, this model is not convenient for simulating real-world networks.

In the context of the social movement of humans, it is needed to use models that have the ability to describe most human mobility characteristics. Hence, Song et al. [42] proposed a model for Individual Mobility (IM) to better describe human movements. Their model is based upon two mechanisms:

Exploration: the tendency to explore new locations decreases with time. Here, the next step is completely independent of the previously visited locations.

Preferential Return: in random walk, the probability of visiting a location is uniform and random. However, human behaviour reflects an important property, that is, the tendency to return to the most visited locations in the past (e.g., home, work).

As an example of how individuals behave in their movements, assume that an individual is positioned at a preferred position at time t = 0. After a waiting time Δt that is selected from the probability P(Δt), the individual changes his/her position. At this point, two options are available:

1. Exploration: the individual moves to a new location with probability:

$$Pnew = \rho S - \gamma, \tag{2}$$

where $\gamma \geq 0$ and $0 < \rho \leq 1$. In this case, the number of previously visited locations S increases incrementally (S = S + 1).

2. Preferential Return: the individual returns to one of the previously   visited locations S with the complementary probability:

$$Pret = 1 - \rho S - \gamma. \tag{3}$$

According to Hugo et al. [43], for individuals, in addition to returning to the most frequently visited locations using the IM model, the probability of returning to the most recent locations is also high. Therefore, Hugo et al. in 2015 proposed a mobility model called Recency-Based Model (RBM), which is basically an extended version of the IM model. The RBM model includes the recently visited locations of individuals in the preferential return mechanism.

NEW CHALLENGES AND TRENDS

In the social framework we described in the Introduction Section, people encounter each other as a part of their social activities, their devices also encounter each other when they become in the communication range of each other (e.g., using WI-FI or Bluetooth technologies). The tracking of encounters requires sensors to memorize their history of encounters. The memory efficiency considerably depends on the size of the sensors' history of encounters. This size is basically dynamic and is expected to gradually increase with time. It can vary from a few items to thousands due to the large number of encounters. This size, in many cases, cannot be handled by the sensors' memory due to strong memory constraints.

To avoid such a case, sensors should use a predefined size of memory and conveniently manage it. This predefined size represents the maximum allowable number of items (i.e., each item represents a history of encounters with a particular sensor) that each sensor can keep in memory. When using a predefined size of memory for sensors, the sensor's memory may become full. This case leads to loss of encounters information that may be important to sensors and eventually changes the pattern of the information dissemination process. This scenario introduces the issue of memory management in sensor networks. To deal with this issue, a decision should be made to replace existing items in a sensor's memory with new and more useful information. This decision should prevent losing information about important encounters of sensors and keep as much of the important encounters information as is possible.

Measuring the strength of ties among network nodes can be considered as the key factor in estimating memory requirements. This is due to the fact that if the approximate number of strong ties (or weak ties) is known, then memory limits can be defined [44]. Measuring the strength of relations among people (including their devices) and estimating memory requirements in WSNs have not received much attention by the research community. Only a few studies have partially investigated the issue of how many friends an individual can have in on-line social networks such as Facebook and Twitter. This, in fact, is due to the availability of the datasets for on-line social networks. There is an existing lack of real datasets for other networks.

Robin Dunbar [44] proposed that the number of social ties that an individual can maintain is approximately 150 according to a human's brain ability. This number is called Dunbar's Number. As mentioned, when estimating memory requirements in wireless sensor networks, information is required on the number of sensors that each sensor in the network should keep in memory. To this end, it is necessary to distinguish the pairs that should be considered as friends. This can be performed by measuring the strength of ties among network nodes. The process of determining the strength of ties is not an easy task because many parameters can be involved in this process (e.g., encounter frequency, duration of encounters, and regularity of encounters) [45, 46]. For example, friends would meet frequently and their meetings are generally longlasting and regular [47].

Lavelle et al. [48] proposed a metric to measure the strength of the relation between two individuals. In their metric, they include the encounter frequencies at a predefined period of time between every two encounters. Their metric has the ability to determine whether a tie is Familiar (strong) or Stranger (weak).

Bandford et al. [49] proposed an approach to measure a tie strength in small mobile communication networks. Their approach is based on call duration and total call count of users to measure the strength of ties among individuals. They found that the use of these two measurements gives a rough estimate of a tie's strength when involving a small number of users, but that tie strength could be better estimated by adding other variables (e.g., text messages between two users).

Mahmood et al. [50,51] proposed a metric for mesuring the stringth of a tie between two nodes. Their method was based on three important characteristics of human relations and interactions, namely, frequecy of encounters, duration of encounters, and regularity of encounters. Then, they used this metric to estimate the number of strong and weak ties that each sensor can maintain and eventually estimate memory requirements.

Furthermore, the same authors proposed a strategy for memory management for each sensor node. Their strategy was based on two social concepts: Social Capital and Preferential Returne mechanism in human mobility.

CONCLUSION

As we see in the description of this chapter, the two fields; WSNs and Social Networks are intertwined. The characteristics of social relations and interactions among people can be used in desiging efficient approaches and methods for WSNs. They also can be considered as a useful tool to overcome the challenges introduced in this area of research. For example, these characteristics are usefull in estimating memory requirements and can be considered as a key factor in developing strategies for memory management in sensor networks. However, other issues and challenges such as sensors connectivity and security need for a more attention by the research community taking into consideration the social characteristics of human when designing approaches.

REFERENCES

[1] Steven B Andrews and David Knoke. Networks in and around organizations. Jai Press, 1999.

[2] David Easley and J Kleinberg. Strong and weak ties. Networks, Crowds, and Markets: Reasoning about a Highly Connected World, pages 47–84, 2010

[3] Mark S Granovetter. The strength of weak ties. American journal of sociology, pages 1360–1380, 1973.

[4] Robert D. Putnam. Bowling alone: The collapse and revival of american community. In Proceedings of the 2000 ACM Conference on Computer Supported Cooperative Work, CSCW '00, pages 357–, New York, NY, USA, 2000. ACM.

[5] S. Jardosh and P. Ranjan. A survey: Topology control for wireless sensor networks. In Signal Processing, Communications and Networking, 2008. ICSCN '08. International Conference on, pages 422 –427, Jan. 2008.

[6] Xiaofang Li, Yingchi Mao, and Yi Liang. A survey on topology control in wireless sensor networks. In Control, Automation, Robotics and Vision, 2008. ICARCV 2008. 10th International Conference on, pages 251 –255, dec. 2008.

[7] Kai Lin, Chin-Feng Lai, Xingang Liu, and Xin Guan. Energy efficiency routing with node compromised resistance in wireless sensor networks. Mob. Netw. Appl., 17(1): 75–89, February 2012.

[8] Kieu-Ha Phung, Tien Pham, Huy Tran, Anh Truong, Lan Tran, and K. Steenhaut. Performance study of link estimation in static wireless sensor networks. In Advanced Technologies for Communications (ATC), 2012 International Conference on, pages 251–254, 2012.

[9] Mario Di Francesco, Sajal K. Das, and Giuseppe Anastasi. Data collection in wireless sensor networks with mobile elements: A survey. ACM Trans. Sen. Netw., 8(1):7:1–7:31, August 2011.

[10]N.A. Pantazis, S.A. Nikolidakis, and D.D. Vergados. Energy-

efficient routing protocols in wireless sensor networks: A survey. Communications Surveys Tutorials, IEEE, 15(2):551–591, 2013.

[11] Theofanis P Lambrou and Christos G Panayiotou. A survey on routing techniques supporting mobility in sensor networks. In Mobile Ad-hoc and Sensor Networks, 2009. MSN'09. 5th International Conference on, pages 78–85. IEEE, 2009.

[12] Jamal N Al-Karaki and Ahmed E Kamal. Routing techniques in wireless sensor networks: a survey. Wireless Communications, IEEE, 11(6):6–28, 2004.

[13] Mohammad Hossein Anisi, Abdul Hanan Abdullah, and Shukor Abd Razak. Energy-efficient and reliable data delivery in wireless sensor networks. Wirel. Netw., 19(4):495–505, May 2013.

[14] Amin Vahdat, David Becker, et al. Epidemic routing for partially connected ad hoc networks. Technical report, Technical Report CS-200006, Duke University, 2000.

[15] Anders Lindgren, Avri Doria, and Olov Schel en. Probabilistic routing in intermittently connected networks. ACM SIGMOBILE Mobile Computing and Communications Review, 7(3):19–20, 2003.

[16] Thrasyvoulos Spyropoulos, Konstantinos Psounis, and Cauligi S Raghavendra. Spray and wait: an efficient routing scheme for intermittently connected mobile networks. In Proceedings of the 2005 ACM SIGCOMM workshop on Delay-tolerant networking, pages 252–259. ACM, 2005.

[17] Curt Schurgers and Mani B Srivastava. Energy efficient routing in wireless sensor networks. In Military Communications Conference, 2001. MILCOM 2001. Communications for Network-Centric Operations: Creating the Information Force. IEEE, volume 1, pages 357–361. IEEE, 2001.

[18] Linton C Freeman. Social networks and the structure experiment. Research methods in social network analysis, pages 11–40, 1989.

[19] Aaron Clauset, Cosma Rohilla Shalizi, and Mark EJ Newman. Power-law distributions in empirical data. SIAM review, 51(4):661–703, 2009.

[20] Claudio Castellano, Santo Fortunato, and Vittorio Loreto. Statistical physics of social dynamics. Reviews of modern physics, 81(2):591, 2009.

[21]Santo Fortunato. Community detection in graphs. Physics Reports, 486(3):75–174, 2010.

[22]Michelle Girvan and Mark EJ Newman. Community structure in social and biological networks. Proceedings of the National Academy of Sciences, 99(12):7821–7826, 2002.

[23]Jean-Fran cois Rual, Kavitha Venkatesan, Hao, et al. Towards a proteome- scale map of the human protein–protein interaction network. Nature, 437(7062):1173–1178, 2005.

[24]Herminia Ibarra. Homophily and differential returns: Sex differences in network structure and access in an advertising firm. Administrative science quarterly, pages 422–447, 1992.

[25]Peter V Marsden. Homogeneity in confiding relations. Social networks, 10(1):57–76, 1988.

[26]Miller McPherson, Lynn Smith-Lovin, and James M Cook. Birds of a feather: Homophily in social networks. Annual review of sociology, pages 415–444, 2001.

[27]Y Connie Yuan and Geri Gay. Homophily of network ties and bonding and bridging social capital in computer-mediated distributed teams. Journal of Computer-Mediated Communication, 11(4):1062–1084, 2006.

[28]Ove Frank. Transitivity in stochastic graphs and digraphs. Journal of Mathematical Sociology, 7(2):199–213, 1980.

[29]Diego Garlaschelli and Maria I Loffredo. Patterns of link reciprocity in directed networks. Physical Review Letters, 93(26):268701, 2004.

[30]Mark EJ Newman. Who is the best connected scientist? a study of scientific coauthorship networks. In Complex networks, pages 337–370. Springer, 2004.

[31]Reuven Aviv, Zippy Erlich, Gilad Ravid, and Aviva Geva. Network analysis of knowledge construction in asynchronous learning networks. Journal of Asynchronous Learning Networks, 7(3):1–23, 2003.

[32]George Kesidis, Takis Konstantopoulos, and Shashi Phoha. Surveillance coverage of sensor networks under a random mobility strategy. In Sensors, Proceedings, volume 2, pages 961–965. IEEE, 2003.

[33]Guolong Lin, Guevara Noubir, and Rajmohan Rajaraman. Mobility models for ad hoc network simulation. In INFOCOM 2004. Twenty- third AnnualJoint Conference of the IEEE Computer and Communications Societies, volume 1. IEEE, 2004.

[34] Mirco Musolesi and Cecilia Mascolo. Mobility models for systems evaluation. In Middleware for Network Eccentric and Mobile Applications, pages 43–62. Springer, 2009.

[35] Takeyuki Hida. Brownian motion. Springer, 1980.

[36] GM Viswanathan, Frederic Bartumeus, Sergey V Buldyrev, Jordi Catalan, UL Fulco, Shlomo Havlin, MGE Da Luz, ML Lyra, EP Raposo, and H Eugene Stanley. L'evy flight random searches in biological phenomena. Physica A: Statistical Mechanics and Its Applications, 314(1):208–213, 2002.

[37] David B Johnson and David A Maltz. Dynamic source routing in ad hoc wireless networks. Kluwer International Series in Engineering and Computer Science, pages 153–179, 1996.

[38] Josh Broch, David A. Maltz, David B. Johnson, Yih-Chun Hu, and Jorjeta Jetcheva. A performance comparison of multi-hop wireless ad hoc network routing protocols. In Proceedings of the 4th Annual ACM/IEEE International Conference on Mobile Computing and Networking, MobiCom '98, pages 85–97, New York, NY, USA, 1998. ACM.

[39] A. V Chechkin, R. Metzler, J. Klafter, and V. Yu Gonchar. "Introduction to the theory of l'evy flights. Anomalous Transport: Foundations and Applications", pages 129–162, 2008.

[40] K. Pearson. The problem of the random walk. Nature, 72(1865): 294, 1905.

[41] F. Bai and A. Helmy. A survey of mobility models. Wireless Adhoc Networks. University of Southern California, USA, 206, 2004.

[42] C. Song, T. Koren, P. Wang, and A. Barabasi. Modelling the scaling properties of human mobility. Nature Physics, 6(10):818–823, 2010.

[43] H. Barbosa, F. Buarque de Lima Neto, A. Evsukoff, and R. Menezes. The Effect of Recency to Human Mobility. ArXiv e-prints, April 2015.

[44] R. IM Dunbar. Neocortex size as a constraint on group size in primates. Journal of Human Evolution, 22(6):469–493, 1992.

[45] S. Moon and A. Helmy. Spectral analysis of periodicity and regularity for mobile encounters in delay tolerant networks. SIGMOBILE Mob. Comput. Commun. Rev., 14(4):37–39, November 2010.

[46] S. Moon and A. Helmy. Understanding periodicity and regularity of nodal encounters in mobile networks: A spectral

analysis. In Global Telecommunications Conference (GLOBECOM 2010), 2010 IEEE, pages 1–5. IEEE, 2010.

[47] E. Bulut and B. K Szymanski. Friendship based routing in delay tolerant mobile social networks. In Global Telecommunications Conference (GLOBECOM 2010), 2010 IEEE, pages 1–5. IEEE, 2010.

[48] B. Lavelle, D. Byrne, C. Gurrin, A. F. Smeaton, and G. JF Jones. Bluetooth familiarity: Methods of calculation, applications and limitations. Mobile Interaction with the Real World, Workshop, 9 September 2007, Singapore, 2007.

[49] J. Banford, A. McDiarmid, and J. Irvine. Estimating the strength of ties in communication networks with a small number of users. In Wireless and Mobile Communications (ICWMC), 2010 6th International Conference on, pages 191–195, Sept 2010.

[50] B. Mahmood, R. Menezes, "A Social-Based Strategy for Memory Management in Sensor Networks in Proceedings of the 5th International Conference on Sensor Networks (SENSORNETS 2016), Italy, Rome Feb. 2016.

[51] B. Mahmood, M. Tomasini, R. Menezes, "Estimating Memory Requirements in Wireless Sensor Networks Using Social Tie Strengths", 10th IEEE International Workshop on Practical Issues in Building Sensor Network Applications (IEEE SenseApp 2015), October 26 2015, Clearwater, Fl, USA.

CHAPTER NINE

CELLULAR SYSTEM DRIVING IoT

Ali Al-Sabbagh[1], Ruaa Asabah[1] and Aqiel Almamori[2]

[1] *Department of Electrical and Computer Engineering, Florida Institute of Technology, Melbourne, FL, USA*

[2] *Department of Enginnering and Information Technology, University of Arkansas Little Rock, AR, USA*

ABSTRACT

Nowadays, society has witnessed a paradigm shift in wireless technology, such as smartphones, tablets, GPS, sensors, to mention a few. Yet with the advent of the so-called Internet of Things (IoT), the possibility that these devices are able to communicate with each other becomes true. Many technologies (e.g., Wi-Fi, 4G, Bluetooth, etc.) are incorporated into the IoT framework. These technologies are important insofar as they are the key factors to developing and improving the IoT. This chapter explores the IoT literature in terms of the communication technology involved: 4G-LTE-A. In addition, In this chapter we present an outline of future keys in the current cellular systems for improving the IoT. Furthermore, this work exhibits how the current 4G LTE-A frameworks can contribute to the design of smart cities. This also, presents an overview of about 4G and 5G with their important features and how will drive IoT in future to build smart cities and smart objects. Scientists are concentrates about the advance implementation of these technologies especially the 5G is the term which is not formally used for any current technology and it is still early to provide an exact definition of 5G. We also present an overview of future keys in cellular system to get better IoT such as D2D, M2M, NOMA and Massive MIMO. This work presents the exact steps in 4G and 5G which important towards IoT and the ability to establish the smart cities. Among these steps and challenges are considered as the pieces of the 5G using massive MIMO, new access techniques, millimeter wave (mm Wave) and M2M, D2D communication. The future fifth generation (5G) cellular networks have drawn great attention from scientists and companies around the world.

Keywords: Internet of things (Iota), Cellular System, D2D, M2M, 4G-LTE-A, 5G.

INTRODUCTION

In recent years, there has been a huge interest in smart objects that can connect to the Internet, to share and make big new services to the world like smart homes, mobile health, and for the industrial applications, such as smart grids, efficient transportation, and logistics. The resulting combination system is represented as the Internet of Things (IoT). [1] For that reason, we must have an efficient wireless networking technology with the development of the world, like Wireless Fidelity (WiFi), Long Term Evolution Advanced (LTE-A),5G, Bluetooth,

ZigBee and others. All these networking technologies are used to achieve communication from device to device (D2D), or device to Machine (D2M). [3].

Figure (1) shows the variety of application that can be provided by the internet of things, because of that there must an efficient wireless networking technologies with the development of the world, like Wireless Fidelity (WiFi), Long Term Evolution Advanced (LTE-A),5G 5G, Bluetooth, ZigBee and others. All these network technologies are used to achieve the communication from the devise to devise (D2D) or the devise to Machine (D2M) perspectives.

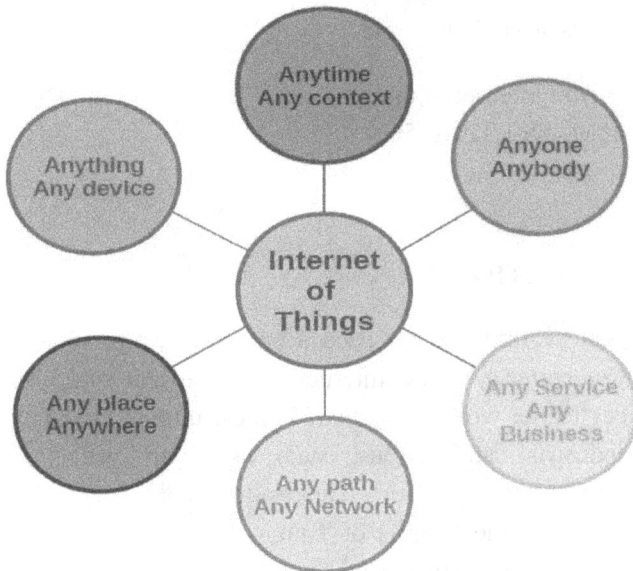

Figure (1). Internet of Things. [1]

Because of the different type of the network equipment another communication between Machine-to-Machine (M2M) also utilized in many circumstances for example inside homes, commercial buildings, schools, hospitals, and factories. [4]

Device-to-device (D2D), the term device refers to the user who uses cell phones or other devices. The communications of D2D is the technics that allow devices to communicate with each other with out the need to access points and the involvement of wireless operators or base stations. Machine-to-Machine (M2M) communications describe algorithms, technologies, and mechanisms which enable wireless and/or wired services and networked devices to exchange information or control data easily, without the need of human intervention. In this case a machine is a device or piece of software.

M2M communications will revolute the telecommunication system operators business because of emergence of new networked applications, this will make many new clients and a huge data flowing in networks. Long Term Evolution Advance (LTE-A) Release 10, is a thoughtful technology produced by the Third Generation Partnership Project (3GPP) it is used in hundreds of networks around the world in many regional or national coverage.

LTE-A is efficient modulation coding schemes, mandatory multi-antenna operation, and very flexible deployment in UHF frequencies are all take in consideration to characterized also the varies type of cell sizes ranging from several kilometers (macro-cells) to few tens of meters (femto-cells) for in-house base stations, called Evolved Node B (eNB). IoT applications are thought by telecom and service operators to be the opportunity to the new generation of the future because of its bight future and advantages it bring pervasive diffusion [5,6].

SMART CITY

This section is very related to the future of IoT applications. And a few researches pay attention to give some details about it in internet of things surveys. Smart City is a term going to be hearing frequently in next years for more details you can back to the chapter of smart city and smart home. The model most commonly adopted to bring business which grows both software and hardware applications for the Internet of Things. This model is represented by figure (2) [3, 5]. In this section I will give some examples about smart cities. The first example is Glasgow, UK. The government has offered ($37 million) for technology that will make the city "smarter, and more comfortable". The new generation of information technology and knowledge-based economy, based on the plan of future cellular network 4G and 5G, this combination of the Internet, and other sensors or devices where Internet of Things technology (IoT).

The second one is Copenhagen in Denmark. The third one California because several companies make the business there. The main features of a smart city and applications are following below [6, 7]:

    1- Smart Lighting: the optimization of street lighting in a better way. Smart Parking: this is a big issue in some countries like the USA. To do the monitoring of the parking availability in the city.

    2- Traffic Congestion: especially in big populated cities. This mission to monitor the vehicles and pedestrian levels. Also Intelligent Highways with warning messages and diversions according to climate conditions.

3- Automation of Public Buildings: Another important application of Iota technology is the monitoring of energy consumption (schools, offices, and museums) by many types of sensors and actuators.

4- Structural health: Monitoring Buildings, bridges and historical sites to evaluate vibration and material conditions.

5- Waste Management: This makes detection of levels in containers to optimize trash collection.

6- Forest Fire Detection and Air pollution: Monitoring of fire conditions to define alert zones. In addition to monitoring the pollution emitted by cars and toxic gases generated in forms. [7,8]

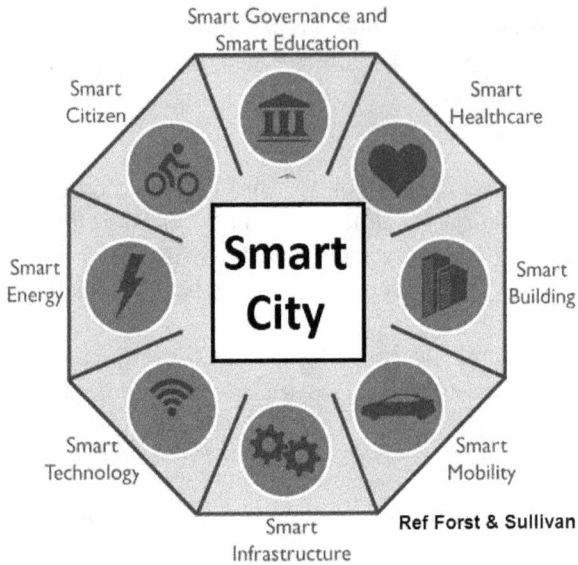

Figure (2). Smart City features.

APPLICATION-1 (CELLOPHANES BASE TECHNOLOGIES)

In this section author will give some brief application for smart parking techniques. These technologies do not need any kind of physical infrastructure in order to function. They are the cheaper alternatives for parking. Some researcher called Lan et al uses in his research GPS, accelerometer and gyroscope sensors detect the pattern of the user's movement in order to infer where they are walking or driving. Moreover, it utilizes techniques such as map matching, which mainly checks if their movement is indoor or outdoor. By modeling the length of user's steps with help of height and speed, their model determines the mode and path of a user's travel. A research Xu et al (2013) employs a similar approach in estimating the availability of the parking spot.

They essentially use trained model to estimate the mode of travel and with the help of patterns of mood changes, the parking spaces are estimated. However, these systems have some disadvantages, such as reliability of data, and the privacy concerns of making the GPS data available to a third party. There are certain non-intrusive cellphone based parking detection solutions like Roadify, as these are more focused on the crowdsourcing part of it. That is the driver has to pick the choice that they have parked a car at a spot in order for the system to sense. However, depending on a large number of drivers makes the reliability part of it more problematic as not everyone might select to report their parking status. [8, 9]

APPLICATION-2 (RFID BASED)

Basically, smart parking is one of the challenges in several countries and will affect apparently in economic and gas consumption also the road traffic, these systems use Radio-frequency identification (RFID) technologies to identify the parking lots available in the area. This is achieved by storing information about the id for the vehicle on a microchip with an antenna. RFID tag, as these are normally known to help identify the entering or leaving vehicles. Overhead scanners detect the information on these tags to record the parking status. Normally three types of RFIDs are used: 1) Active, 2) Passive and 3) semi-passive. Active tags powered with own source and are considered to be better functioning. RFIDs a better choice in practice, as unlike barcodes multiple RFIDs can be read at the same time, reducing the time of operation.

However, due to their low price, passive tags are more common in parking management systems. A research study by Ganesan et al. (2007) after applying RFID to collect information about the availability of parking, it updates to the web servers continually to help the drivers. The Ganesan discuss an automated parking space allocation system by splitting this task into 4 modules, 1) Serial Port Communicator, 2) Free slot checker, 3) Parking charge calculator and 4) Free slot viewer. It provides a functionality where drivers can communicate with the system to get results in the form of a text message and they will also be able to park in a space with the help of a specific text message. [8, 9]

INTERNET OF THINGS RÉQUIREMENTS

There are three important requirements of IoT Sensing data, Processing and connectivity, all these requirements create an obstacle to implement IoT in the future by challenging us on several issues related to the huge number of devices and huge data transferred around the world.

Internet passed some phases until receiving IoT, starting from PC-PC connection [10], then step by step to a higher level of cloud to arrive to IoT as shown in the figure below:

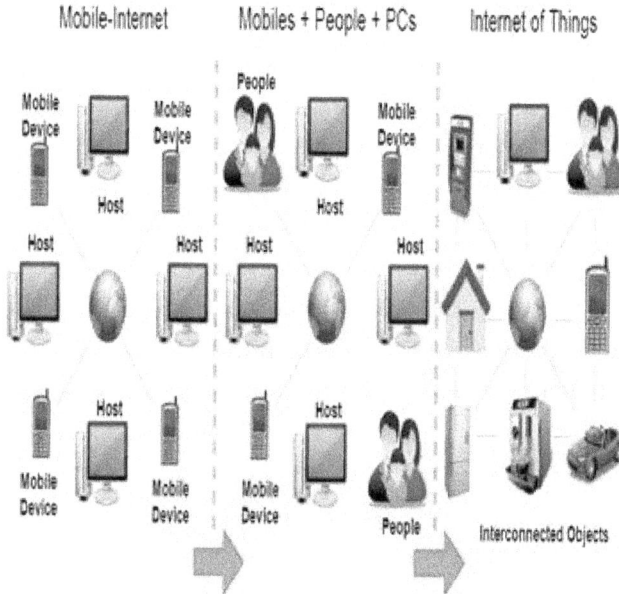

Figure (3). Internet stages to Iota [2]

The challenges of IoT-devices can simply clarify as follow [11, 12]:

1- Signaling:

A reliable bidirectional signaling is very important with IoT connected devices to make the routing of data easy. Devices may be talking to a server to collect data, or the server may be talking to the devices, or maybe those devices are talking to one another. In other word data needs to sent, and received back from point A to point B in a fast and secure way.

2- Security:

Security is a huge umbrella, but it is paramount in the Internet of Things connectivity. This can simply follow an authorization: When sending or receiving a stream of data, it's essential to make sure that the IoT device or server has authorization to send or receive that stream of data.

3- Open ports:

An IoT device is dangerously defenseless when it's listening to an open port out to the Internet. Then Encryption is needed (end to end encryption) between IoT devices.

4- Presence Detection:

By knowing immediately when an IoT device drops off the network and goes offline, or come back online. Presence detection of IoT devices gives an exact, up to the second state of all devices on a network. This gives the ability to monitor IoT devices and fix any problems that may arise with your network.

5- Power consumption:

Sending data between one another (IoT Devices) takes a toll on power and CPU consumption. With all this communication we need an efficient cellular system using HetNet 4G or 5G as a long battery life, also a smart sensor built in IoT devices and figure below shown the relation between receiving power versus the distance by HetNet.

6- Bandwidth:

As expected in the future, thousands of IoT devices are connected to the network, so bandwidth consumption is yet another challenge for IoT connectivity. Bandwidth on a cellular network is expensive, that's why we discussed LSA in 5G section. [2, 11, 12]

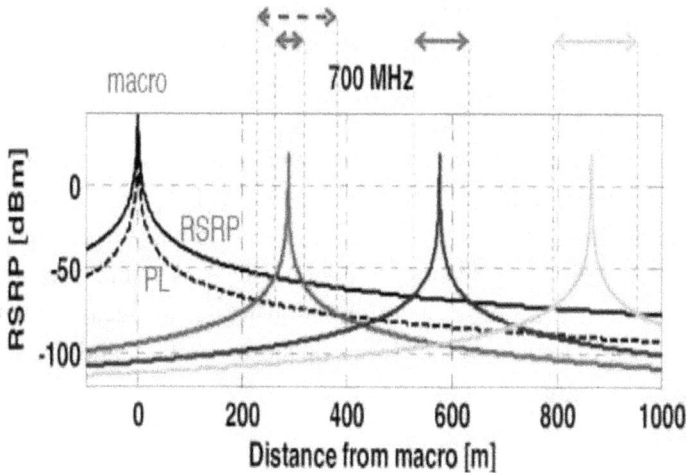

Figure (4). Receive signal VS distance.

INTERNET OF THINGS PRÉDICTIONS

In the next five years, IoT and the Cloud of more than 90% of all IoT data will be hosted on service provider platforms as cloud computing reduces the difficulty of assisting IoT "Data Blending". Also within three years network capacity, 50% of IT networks will transition from having excess capacity to handle the extra IoT devices, to being network constrained with closely

10% of sites being overwhelmed. By the end of 2018, 40% of IoT-created data will be stored, processed, and acted upon close to, or at the edge, of the network. Today, over 50% of IoT activity is centered in manufacturing, transportation, smart city, and consumer applications. With the Challenges to build innovative and sustainable smart cities, local governments will represent more than 25% of all government external spending to organize, the commercial value of the IoT by 2018. Within five years, 40% of wearables will have evolved into a viable consumer mass market. By 2018, 16% of the population will be Millennial and will be accelerating IoT adoption due to their reality of living. IoT will become open-sourced, allowing a rush of vertical-driven IoT markets to form IoT and embedded systems. By 2018, 60% of IT solutions originally developed as branded, closed-industry solutions. This section we have discussed some statistical information to get an idea about the future and challenges and some of these numbers and facts are shown in Figure (5): [13, 14]

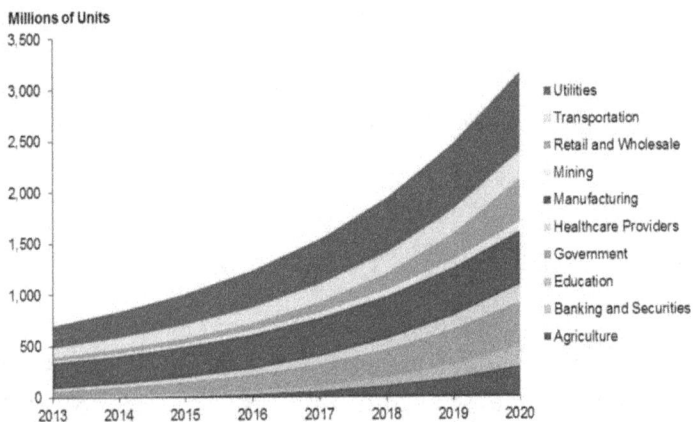

Figure (5). Expected Iota devices -Year 2020. [2]

4G LTA-A Improvisent for IoT

Because of the trends and problems for developing the Internet of Things (IoT) concept, a very challenging task is to connect thousands of devices into one large network. First of all, we have to understand why everybody is interested in this new development. The main purpose is by the creation of a new class of wireless applications. The devices will be different, such as, smart fridge, that have the ability to transmit information about the quality of food inside at the central device, another example could be smart-phone in a smart house grid and to a sensor measuring pressure or temperature and sending this data to a central decision-making controller unit onto the automated factory. Now we could derive the main achievement of any IoT network: machine-based devices as network units, high number of devices in the network and centralized data. These main properties will lead to several implementation tasks. The LTE radio access network should be enhanced to fit with the need of IoT communications many studies by 3GPP are going to make that possible. These developments consider many features such as issues of network support for IoT devices the overload control, power saving for ultra-long battery life, device cost reduction, coverage enhancement and signaling overhead. The process of standardizing new features is done by 3GPP in Table (I). The features listed in Table (1) ensure that LTE can achieve IoT requirements of ubiquitous coverage, ultra-long battery life and low-cost devices. To support machine traffic and co-existence of humans, service differentiation of the various techniques can handle different traffic types [15, 16].

TABLE (1)

LTE Release	Feature
Rel-11 (2012)	• UE power preference indication • RAN overload control
Rel-12 (2014)	• Low-cost UE category (Cat-0) • Power saving mode for UE • UE assistance information for eNB parameter tuning
Rel-13 (expected 2016)	• Low-cost UE category • Coverage enhancement • Power saving enhancement

Some of them are: time-controlled access of M2M devices, including access grant time, forbidden time interval, and limiting services to M2M devices if their behavior is not aligned with M2M features. Scheduling prioritization and semi-persistent scheduling to lower overhead can also be used. Another feature that exists in the LTE-A is Cell Splitting, this can take advantage of the backward use of the spectrum which is one of the main problems. In Cell Splitting the Mobile communications are rapidly developed so that the number of users of cellular networks is increasing. At the same time Wi-Fi networks are also serving an increased amount of users. On the other hand the continuous interest to mobile users in the fourth generation of cellular communication - LTE.

This is due to low cost of service of mobile telecommunication and free Internet access with using WLAN ad-hoc. In the traditional cellular concept, in order to improve capacity of a network, the number of base stations deployed should be increased. But this will worsen the interference, which means that the average cell radius should be decreased. None-the- less, this will lead to a new situation where the size of signaling of mobile users will dramatically increase because of the growth of cell edge reconnections from the tiny base station coverage. Some solutions in such conditions do not work. At the same time, many other alternatives that operate concurrently by the progress in mobile communication networks have been created, such as HetNets and D2D. By Combining WLAN with cellular networks, a significant increase in the total capacity of the networks can be provided, this is at the core of the concept of HetNets [17, 18].

In HetNet, the concept of a base station extends from the conventional cellular network and includes macro, micro, pico and femto-cell. Usually the range of macro is 1km-5km, for micro 100m-500m a pico is 10-50 meters or less, and a femto is less than 10 meters. Table (2) shows the numbers of the transmit power and the coverage and the backhaul of each type. They have great potential for co-development with cellular networks. All these elements in this network are part of a heterogeneous network and they communicate with each other.

Types of nodes	Transmit power	Coverage	Backhaul
Macrocell	46 dBm	Few km	S1 interface
Picocell	23–30 dBm	< 300 m	X2 interface
Femtocell	< 23 dBm	< 50 m	Internet IP
Relay	30 dBm	300 m	Wireless
RRH	46 dBm	Few km	Fiber

TABLE (2)

An effective use of resources is allowed by this feature for optimizing the network. The structure of HetNet's cell is shown in Figure (6). The cell is served by a base station LTE network (macro BS), which has a number of WLAN access points that are now part of the global network. In the necessity case the users can use a resource of the cellular or Wi-Fi network. In addition, the evolution of a technology to another is automatic. For instance if we have static clients who are in the same place for a long time and use a packet data connection, it is recommended to change their traffic to Wi-Fi. At the same time, mobile users quickly through the cell's border, it would be best to use the resources of macro BS for them.

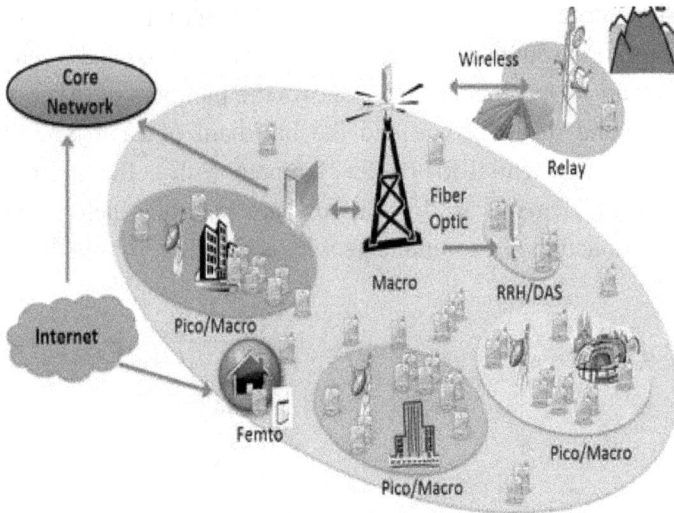

Figure (6). The structure of HetNet's cell. [2]

D2D COMMUNICATION IN LTE-A

One of the achievements of 3GPP Release 12 is the development of LTE (D2D) discovery and device communication with the device. Release 12 is just the beginning, but over time D2D will create a number of new service opportunities, at the same time achieving the performance and efficiency of important advantages in LTE networks. To do this, allowing mobile devices to detect the presence of other devices in the vicinity and communicate directly with them, with minimal involvement from the network. LTE D2D is a complex development, but a great opportunity for mobile network operators; it can have serious consequences for the design of the mobile network.

D2D set a precedent for networks to relinquish some of their control over mobile devices and traffic and that will have an impact on existing services and future planning of the network. During these years, an important area is being interested for the especially social networks and Internet industry, which can achieve by the proximity-based information and applications The most important application can be made to a user at a given time in a particular place at a particular time situation, the more valuable it becomes, and the more likely that the user will act on it [16, 19]. A secure, dynamic, flexible, efficient a decentralized approach to proximity discovery and device-to-device (D2D) communication, is provided by enabling proximity-based services to flourish LTE D2D. The 3GPP solution for D2D is termed Proximity Services (PROSE) and there are two components, as shown in figure below:

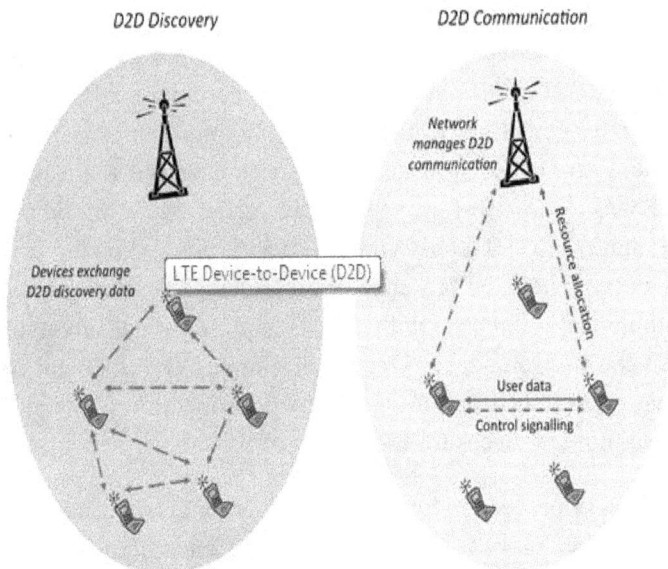

Figure (7). Main structure of discovery and direct communication

The first is D2D Discovery, to allow the mobile device within network coverage use of the LTE radio interface in order to discover the presence of other D2D-capable devices in the nearest region and, where permitted, to share certain information with them. And it is used for public safety and commercial. The second one is D2D Communication, which is the ability of D2D mobile devices for using the LTE radio interface to communicate directly with each other so there is no need for routing the traffic through the LTE network. The network exerts with the control of the radio resource allocation and security of the connections. Table 3 shows briefly the difference between the discovery and direct communication.

TABLE (3) [2]

D2D Types: Discovery and Communication	Within network coverage	Outside network coverage
D2D Discovery	Public safety and commercial	None
D2D Communication	Public safety and commercial	Public safety only

There are some benefits from direct communication of devices that can be provided to users in different applications when the devices are in close proximity:

A- *Use of licensed spectrum:* LTE would use licensed spectrum, unlike other device to device systems such as Wi-Fi, Bluetooth, and this would allow the frequencies to be used in a way that make it subject to less interference, thereby allowing the flexible communication.

B- *Data rates:* Devices may not have the ability to support high data rate transmission because they might be remote from cellular infrastructure, and that data rate may be required.

C- *Instant communications:* The devices could be used for instant communication between a different numbers of devices in the same way that walkie-talkies are used because the D2D communication does not rely on the network infrastructure. This is especially used in emergency services.

D- ***Reliable communications:*** To provide good, reliable communication, LTE Device to Device can be used to communicate locally between devices such as in the case of the field of LTE network for any reason – even in the result of a disaster.

E- ***Interference reduction:*** To reduce the overall level of interference, fewer links are needed by not having to communicate directly with a base station, (i.e. essentially only between devices) and this has an effect on the amount of data being transmitted within a given spectrum allocation.

F- ***Power saving:*** Energy saved by the use of D2D communication provides for many reasons. For example, if there are two devices that are in close proximity then lower transmission power levels are required. [18,19]

M2M COMMUNICATION IN LTE-A

The communication of machine to machine can be enrolled in many branches in our life, such as, safety and protection, smart home technology, e-health, sensor networks in factories, education is an incomplete list of areas in which machine-to-machine communications could be achieved, as shown in figure (8). M2M is a generalized growth of mobile communications with a host of applications, in the future. The 3rd Generation Partnership Project (3GPP) will standardize the principle of M2M application operation in 3GPP networks. Because of M2M communication happening between different machines (sensors usually) and the core network. We will have about 300 billion devices with M2M trend. M2M is used in navigation, security systems, communication between different objects, health systems, etc. These devices should meet certain requirements based on the operating conditions [5, 6]:

A- The first is by having low energy consumption for data transmission; three states could be in each device: idle, transmit and receive. At any period of time a sensor transmits once in the common mode. The transmission frequency is selected depending on working conditions and the environment such as (1 min., 5 min., 1 h., Etc.). For example, a video camera may be transmitted once in 5 min, so that the data size does not exceed several bytes. It may be information on the pressure, temperature, etc. Energy device should be minimal, since the data transmitted frequently and in small portions. In addition, a great loss of energy leads to a brief operation of each device. As the lifetime of the battery becomes an important factor in their design. For example, if the sensors are working under extreme conditions during a long period of time, interference will happen between the high cost of energy and data collection.

B- More simplicity; there is a different variety of M2M devices, it might be from tens to tens of thousands. In a single cell the number of devices may grow up to 30,000. The complexity of the device affects the price as well, because high hardware complexity is very expensive. It would not be feasible for an enterprise to install several thousand units with the high cost per unit.

C- Long battery life; it is necessary to have long battery life for the sensor to provide sufficient energy capacity which will help it run for a long time. There are important factors that play a particularly important role in many conditions where access is limited to the detection and the charging capacity is minimal. Also advances in battery development to create a device that can function independently for a long time. When using an M2M device in extreme conditions, the operating time is an important factor. [15,18,19]

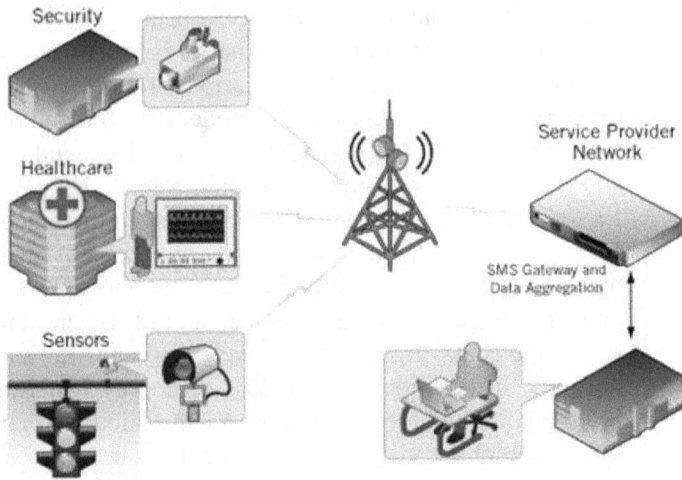

Figure (8). M2M different application. [2]

FIFTH GENERATION (5G) TECHNOLOGY

The 5G is the next generation for mobile system proposed to be commercially 2020 this technology has amazing data capabilities and high data broadcast within the latest mobile operating system. 5G provides high connectivity and universal coverage by using the same multitier network in 4G. Some of its features are following: [20]

1) 5G Cellular system provides a high resolution for bi-directional large bandwidth with effective billing system.

2) 5G Cellular system offers subscriber supervision tool.

3) High quality service by using special policy to avoid the error.

4) 5G Cellular system provides large data in terms of gigabit.

And will become more accurate.

5) 5G provides up to 25Mbps connectivity speed. And uploading, downloading speed is touching peak. [21] This will come by using the splitting techniques from 4G with new access scenarios by Macro, Micro, Pico, relay and Femto to get more users in small coverage areas as shown in the figure:

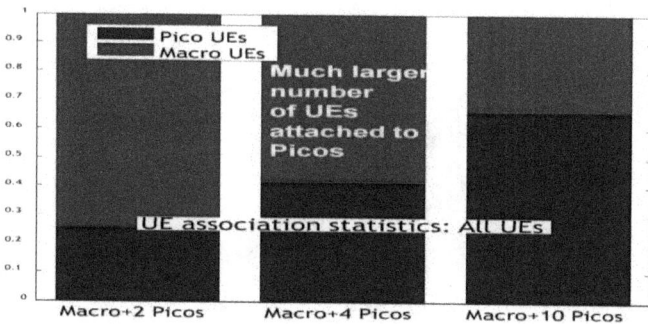

Figure (9). (Macro + Pico) attached users.

MASSIVE MIMO

Massive MIMO is the solution to access thousands of arrays to be configured at the base station. For more accurate beam control and greater spectral efficiency than legacy MIMO, in legacy base station has an eight antenna arrays. But massive MIMO also involves the use of spatial multiplexing and interference mitigation to increase system capacity. Some research work to get more than 64 antenna arrays. This required a new antenna design, and reference signal design still needs additional effort. This called CoMP which is mean coordinated multi-point transmission. Massive MIMO is an extension of CoMP. this is clearly shown in the figure [20, 21] Massive MIMO relies highly on a property of the radio environment called favorable propagation, meaning that the propagation channel responses from the base station to active terminals are sufficiently different

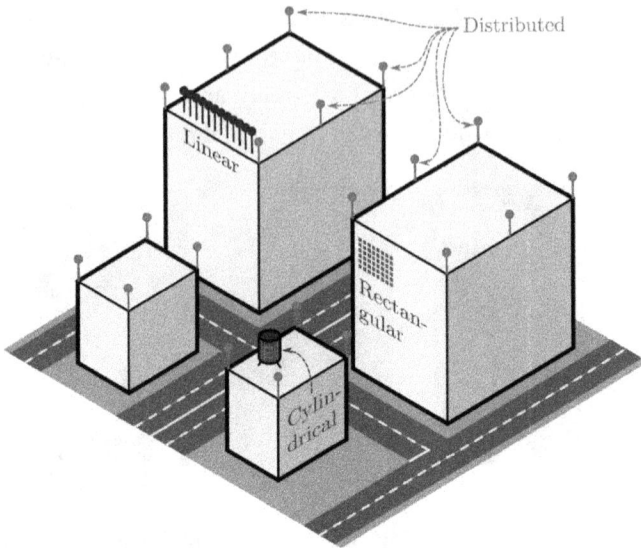

Figure (10). 5G-Massive MIMO. [22]

Massive MIMO is also called Large Antenna System (LAS), which is a promising technique to increase the spectral efficiency of cellular networks. Massive MIMO systems deploy antenna arrays with hundreds or more of active elements at the base stations and perform coherent transceiver processing. A possible configuration and deployment scenario for a massive MIMO base station is shown in Figure (10). [22] It is regarded as an efficient and scalable approach for multicell multiuser MIMO implementation.

By deploying base stations (BSs) with much many more antennas than active user terminals, it is possible to attain the asymptotic orthogonality condition among MIMO channels and manage and mitigate intra- and inter-cell interference. Hence, using simple linear precoder and detector we can approach the optimal dirty-paper coding capacity (DPC) [23, 24]. Channel state information (CSI) at BSs have a central role in the exploitation of channel orthogonality. The uplink training sequence has the advantages of the length of the training sequences is proportional to the number of active terminals rather than that of BS antennas. The training length is limited by the channel coherence time, which can be short due to terminals of high mobility. It has been shown that the impact on channel estimation due to the use of short training sequences can be minimal in the LAS regime, specifically, among the poorly estimated channels, and that the asymptotic orthogonality condition can still hold [25]. However, a major problem with time-division duplex (TDD) massive MIMO is the reuse of the same pilot sequences in multiple cells. The channels to terminals in different cells that share the same pilot sequence will be collectively picked by BSs. That is, the desired channel learned by a BS is contaminated by undesired channels. If this contaminated channel state information (CSI) is used for transmitting or receiving signals, intercell interference occurs directly, limiting the achievable signal-to-interference-noise ratio (SINR). This phenomenon, known as pilot contamination, cannot be avoided simply by increasing the BS antenna size. [26, 27]

More specifically, a set of orthogonal pilot sequences is assigned to the users within a cell that are transmitted during specific time slots. Assume that in any transmission-reception phase, τ time slots are assigned to the pilot sequences. As the number of orthogonal τ-tuples is τ, then, there are a limited number of orthogonal pilot sequences. Therefore, it is essential to reuse the same pilots in different cells. or even in the same cell. Consequently, the received signals at each BS is corrupted by the nterference from th same assigned pilots originated by different users. The effect of pilot contamination is usually quantified as SINR saturation due to intercell interference. Many papers have studied this saturation phenomenon and the corresponding uplink or downlink throughput. In other hand the TDD mode, the links occupy the same frequency band but are allocated different slots in time, while in FDD they occupy different frequency bands. In the uplink of both TDD and FDD, each terminal transmits pilot symbols that are orthogonal to other terminals within the cell. TDD offers a distinct advantage: the estimates in the uplink are valid for the downlink as well. In FDD, the BS needs to find downlink estimates separately, which increases the channel estimation overhead. In FDD mode, each terminal reports the M ×1 channel vector over the reverse link, which consumes additional resources proportional to M, thereby increasing the latency M-fold before the BS can use them for multiplexing in the downlink. Therefore, in a LAS system, FDD is generally infeasible using the pilot-based training schemes. How different the channel responses are to different terminals is determined by how much the spread is between the smallest and largest Eigenvalues of the channel responses matrix. [28]

NEW MULTIPLE ACCESS AND MILLIMETER WAVE

NOMA is non-orthogonal multiple access which is a new proposed solution instead of OFDM using non orthogonal signal. The information at the transmitter side is superposed in the power domain or code domain and is demodulated at the receiver side. NOMA also can be used in NFC near-far between users and obtain the maximum sum rate. NOMA can be considered as power-domain or code-domain. The first one NOMA power domain is an extension of the spatial, time or frequency domains and enables more users to access in the limited case. The second one NOMA Code-domain contains low-density CDMA and interleave division multiple access (IDMA) this is a combination idea. NOMA is mostly used high dense area, or access cannot be easily synchronized like D2D Discovery phase. Or can be used in different case which is the base station has a comparatively small number of antennas not Massive MIMO. Such an ultra-dense networks, D2D communication or M2M communication. [29, 30] Also, CR: Cognitive radio was often pushed as a solution to the problem of frequency spectrum shortage; it is rarely adopted as there are always concerns about the impact on the primary user or license holder of the spectrum. Nowadays there is a vision for alternative solution proposed which can seriously solve this issue is (ASA): Authorized Spectrum Access or some time called (LSA): Licensed Spectrum Access. Simply the idea of LSA is to permit approved users to access licensed spectrum based on conditions set by the licensee of the spectrum. This allows the spectrum to be more efficiently used and solve the problem of QoS: quality of service for the users. [30,31]

Basically the bandwidth is an important issue for high throughput, by using a millimeter wave length will go higher and higher in frequency not like the current one in 4G up to 6 GHz. That why there is effort to look beyond 6 GHz up to 60 GHz and also at the millimeter wave frequencies to evaluate their feasibility for use in future networks. But we have a challenge for outdoor cases because broken waves in buildings, at the same time in future have to use a new channel modeling in different scenarios. This will be required before transmission technologies can be designed for them. Millimeter wave frequencies hold the most promise, and there are already on-going efforts to make this a possibility. This is very usefull for indoor applications with high dense devices with small area. We knew that we have a huge band in millimeter wave; therefore, it hasn't had the same characteristics for all requencies. Each specific sub band has special characteristics. Atmospheric absorption and rain attenuation are the most important issues that we should know about each requency to choose it as a carrier frequency in our application. we understood that why the researchers started with the frequencies 28GHz and 38GHz (they had small loss due to atmospheric absorption). It is about 0.06 dB/Km for 28GHz and about 0.08 dB/Km for 38GHz.

In addition, the frequency ranges from 70 to 100GHz and from 125 to 160GHz also have small loss. Researchers made measurements in New York City for 28 GHz band using horn antenna that has 24.5 dBi gain and 10° half power beamwidth for both transmitting and receiving. Free space path loss was measured for 5-meter distance between transmitter and receiver then the penetration loss of each material was measured by putting the material between transmitter and receiver with the same reference distance (5m). It showed that tinted glass has huge penetration loss about 40.1 dB and brick has penetration loss about 28.3 dB. These are main materials in outdoor.

For indoor materials, the clear glass and wall have less penetration loss about 3.9 dB and 6.8 dB respectively. As a result, higher isolation exist between indoor and outdoor networks.

Another important issue in any transmitted signal is how much Doppler spreading will effect on it? We knew that the Doppler Effect depends on the variation of the channel within the symbol duration so that if the symbol uration is greater than the coherence time the signal will suffer from Doppler spread. Also, we knew that Doppler Effect increases with increasing of frequency and speed of transmitter and/or receiver which means the mm-wave frequency will suffer from Doppler Effect but we decided that this type of frequencies will use with wide BW, which means very small symbol duration (less than coherence time of the channel). It means that yes, there is Doppler Effect but it will be not significant. In addition, we will not observe huge small scale fading because the symbol duration is very small and the number of multipath components are little. [32, 33]

5G IMPROVEMENTS FOR IoT

IoT expected devices is very huge by attached smart sensors to all the objects in our life. A cellular 5G mobile systems will require a combination of new system ideas to boost the spectral and energy in efficient way. This work presents the main requirements for 5G wireless systems to fit IoT are outlined below: [2, 12, 34]

1- **Data rate and latency:** in 5G networks are intended to enable practiced data rate of 300 Mbps and 60 Mbps in downlink and uplink. The end-to- end latencies are estimated to be in the order of 2 to 5 milliseconds.

2- **Base station (BS):** BS densification is an effective methodology to meet the requirements of 5G, according to the large number of low power nodes, relays, and device-to-device (D2D) communication links with much higher density in small area. This is a smooth development of 4G as shown in figure (11) displays a multi-tier network with a macrocell, relays, picocells, femtocells, and D2D links.

3- **Machine-type Communication (MTC) devices:** This step after M2M and D2D as discussed in previous section, which is number of traditional human-centric wireless devices with Internet connectivity such as vehicles, home appliances.

4- **Multiple RATs:** this is important key and the main difference than 4G inside the architecture of 5G which is not about changing the existing technologies, but to enhance and support them with new technologies. The desire 5G systems will continue to evolve to provide a superior system performance. In 4G- RATs, including GSM, HSPA and LTE.

5- **Energy-efficient communication:** One of the main challenges in 5G wireless networks is to improve the energy efficiency of the battery wireless devices. And the combination of different energy harvesting technologies may be essential for macrocell communication, by using (e.g., solar and wind energy). Practical circuits for harvesting energy are not yet available since the conventional receiver architecture is designed for information transfer only.

6- **Millimeter-wave communication:** To satisfy the exponential growth in network traffic and number of devices with their future variety beyond the current 4G standard. and there are already on-going efforts to make this a possibility. This is useful for indoor applications with high dense devices with small area. To be precise the use of millimeter-wave frequency bands (e.g., 28 GHz, 38 GHz or 60 GHz bands) instead of the conventional 20 MHz channels for 4G.

7- **Prioritized spectrum access:** The ideas of both traffic based and tier-based priorities will exist in 5G networks, e.g., reliability and latency requirements, energy constraints.

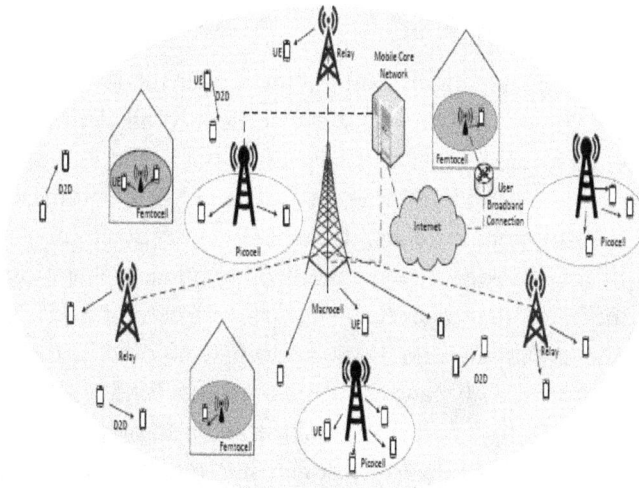

Figure (11). 5G network with multi-tier.

Basically, the macro and femto users play the role of high-Low priority users. In the uplink route, macrocell users at the cell edge typically transmit with high powers, which is cause a high uplink interference to close femtocells. D2D is transmission where different devices may opportunistically access the spectrum to form a communication link between them. A multi-tier network gives an idea about macro cells, Pico cells, Femto cells, relays, and D2D links. Arrows indicate wireless links, whereas the dashed lines denote the backhaul connections. This is an important feature will extend in 5G networks to allow other nodes, rather than the macrocell BS, to have the control. For example, consider a D2D link at the cell edge and the direct link between the D2D transmitters UE to the macrocell is in deep fade, then the solution by using the relay node to control the signaling.

CONCLUSION

The IoT has gained significant attention over the last few years. This idea comes by attached smart sensors to all the objects in our life. Also 5G is the next step after 4G in the evolution of mobile communication after using D2D and M2M and these will be a keys component of IoT to enable connectivity for a wide range of applications beyond those of previous generations by using the same heterogeneous networks of current LTE-A. 4G opens the doors through D2D communication to get smooth change towards 5G because new requirements for 5G are already unleashing a flurry of creative thinking and a sense of urgency in bringing innovative new technologies into reality. To implement an IoT, this requires a massive MIMO system with a new access technique such as NOMA and Licensed Shared Access. Also a mm-Wave cellular system 60GHz was considered something strange but now it is practically considered an predictability. In Future 5G networks will be primarily designed to expand system performance and provide new services by the expected large number of connected device. This technology will increase the capacity and single user data rate, also reducing the delay by increasing the number of terminal connections. , it is a long road ahead to truly IoT and smart cities using 4G then 5G networks. Many technical Challenges remain under research to help pave the road to IoT and 5G.

ACKNOWLEDGEMENT

Our Team greatly appreciate Mr.Nabeel Suleiman, who is a PhD student at University South Florida, Tampa, USA for discussions and valuable comments.

REFERENCES

[1] L. Atzori, A. Iera, and G. Morabito, "The internet of things: A survey", Comput. Netw, vol. 54, no. 15, pp. 2787–2805, 2010.

[2] Ruaa A. Saeed Alsabah, Ahmed Yaseen Mjhool, Ali Abbas Al-Sabbagh, "4G LTE-A Improvements Towards the Evolution of Internet of Things (IoT): Survey", Journal of Networks and Telecommunication Systems, Vol. 1 (1), 11-19, August 2015 ISSN: Pending, DOI: Pending. Published online: www.unitedscholars.net.

[3] Michele Nitti, Luigi Atzori," Friendship Selection in the Social Internet of Things: Challenges and Possible Strategies" IEEE INTERNET OF THINGS JOURNAL, VOL. 2, NO. 3, JUNE 2015.

[4] Fagen Li and Pan Xiong. "Practical Secure Communication for Integrating Wireless Sensor Networks Into the Internet of Things". IEEE SENSORS JOURNAL, VOL. 13, NO. 10, OCTOBER 2013.

[5] Julien Beaudaux, Antoine Gallais, Julien Montavont. "Thorough Empirical Analysis of X-MAC Over a Large Scale Internet of Things Testbed". IEEE SENSORS JOURNAL, VOL. 14, FEBRUARY 2014.

[6] Ali Abbas Al-Sabbagh, Ruaa. A. Saeed.Alsabah, Ahmed.Yaseen. Mjhool, "An extensive review: Internet of things is speeding up the necessity for 5G" Vol. 5 - Issue 7 (July - 2015), International Journal of Engineering Research and Applications (IJERA), ISSN: 2248-9622 , www.ijera.com.

[7] Ruaa Alsabah, Ali Al-Sabbagh, Josko.Zec and I. Kostanic, "Satellite Versus Airborne Techniques", Chapter 5 in the "Advances in Computer Networks and Information Technology, Vol:I", Jan, 2016, ISBN-13: 978-0692618707.

[8] Ahmed Yaseen Mjhool, Ali Abbas Al-Sabbagh, Ruaa A.

Saeed Alsabah, "Smart Parking Techniques Based on Internet of things", Journal of Networks and Telecommunication Systems, Vol. 1 (1), 1-10, August, 2015.

[9] Lan, Kun-Chan, and Wen-Yuah Shih. "An intelligent driver location system for smart parking." Expert Systems with Applications 41.5 (2014): 2443-2456.

[10]Xu, Bo, et al. "Real-time street parking availability estimation." Mobile Data Management (MDM), 2013 IEEE 14th International Conference on. Vol. 1, 2013.

[11]C. Perera, A. Zaslavsky, P. Christen, and D. Georgakopoulos, "Context aware computing for the Internet of Things: A survey," IEEE Commun. Surveys Tuts., vol. 16, no. 1, pp. 414–454, First Quarter 2014.

[12]H. Sundmaeker, P. Guillemin, P. Friess, and S. Woelffle, "Vision and challenges for realising the internet of things," European Commission Information Society and Media, Tech. Rep., March 2010.

[13]Ahmed. Banafa, Openmind-April 2015.

[22]https://www.bbvaopenmind.com/en/internet-of-things-opportunities-and-challenges

[14]J. Beaudaux, A. Gallais, and T. Noel, Heterogeneous MAC Duty-Cycling for Energy-Efficient Internet of Things Deployments. New York, NY, USA: Springer-Verlag, 2013.

[15]Jun Huang, Member, IEEE, Yu Meng, Xuehong Gong, Yanbing Liu. "A Novel Deployment Scheme for Green Internet of Things". IEEE INTERNET OF THINGS JOURNAL, VOL. 1, NO. 2, APRIL 2014

[16]Costantino, Luca, "Performance analysis of an LTE gateway for the IoT."World of Wireless, Mobile and Multimedia Networks (WoWMoM), 2012 IEEE International Symposium on a. IEEE, 2012.

[17]Pereira, Carlos, and Ana Aguiar. "Towards Efficient Mobile M2M Communications: Survey and Open Challenges." Sensors 14.10, 2014.

[18]Ratasuk, Rapeepat, et al. "Recent advancements in M2M communications in 4G networks and evolution towards 5G." Intelligence in Next Generation Networks (ICIN), 2015 18th International Conference on. IEEE, 2015.

[19]Hossain, Ekram, et al. "Evolution toward 5G multi-tier cellular wireless networks: An interference management perspective." Wireless Communications, IEEE21.3 (2014):

118-127.

[20]Dementev, Oleg. "Machine-type communications as part of LTE-advanced technology in beyond-4G networks." Open Innovations Association (FRUCT), 2013 14th Conference of. IEEE, 2013.

[21]Dementev, Oleg. "Machine-type communications as part of LTE-advanced technology in beyond-4G networks." Open Innovations Association (FRUCT), 2013 14th Conference of. IEEE, 2013.

[22]Larsson, Erik, et al. "Massive MIMO for next generation wireless systems."Communications Magazine, IEEE 52.2 (2014): 186-195.

[23]X. Gao, O. Edfors, F. Rusek, and F. Tufvesson, "Linear precoding performance in measured very-large MIMO channels," in IEEE Vehicular Technology Conference, pp. 1–5, Sept. 2011.

[24]Costa, Max HM. "Writing on dirty paper (corresp.)." Information Theory, IEEE Transactions on 29.3 (1983): 439-441.

[25]T. Marzetta, "How much training is required for multiuser MIMO," in Fortieth Asilomar Conference on Signals, Systems and Computers, pp. 359–363, Oct. 2006.

[26]H. Q. Ngo, E. Larsson, and T. Marzetta, "Energy and Spectral Efficiency of Very Large Multiuser MIMO Systems," IEEE Transactions on Communications, vol. 61, pp. 1436–1449, April 2013.

[27]T. Marzetta, "Noncooperative cellular wireless with unlimited numbers of base station antennas," IEEE Transactions on Wireless Communications, vol. 9, pp. 3590–3600, Nov. 2010.

[28]J. Hoydis, S. ten Brink, and M. Debbah, "Massive MIMO in the UL/DL of Cellular Networks: How Many Antennas Do We Need?" IEEE Journal on Selected Areas in Communications, vol. 31, pp. 160–171, Feb. 2013.

[29]D. Astely, E. Dahlman, G. Fodor, S. Parkvall, and J. Sachs, "LTE released beyond," IEEE Commun. Mag., vol. 51, pp. 154–160, July 2013.

[30]Chandra S. Bontu, Shalini Periyalwar, and Mark Pecen Wireless Wide-Area Networks for Internet of Things. IEEE vehicular technology. March. 2014.

[31]Luigi Atzori. Antonio Iera. "Smart Objects" to "Social Objects": The Next Evolutionary Step of the Internet of Things

IEEE Communications Magazine January, 2014.

[32]T. Rappaport et al., "Millimeter Wave Mobile Communications for 5G Cellular: It Will Work," IEEE Access, May 2013.

[33]H. Zhao et al., "28 GHz Millimeter Wave Cellular Communication Measurements for Reflection and Penetration Loss in and around Buildings in New York City," IEEE ICC, 2013.

[34]Rita C. Nilawar1, D.M. Bhalerao. "REVIEW ON A NEW GENERATION WIRELESS MOBILE NETWORK 5G". IJRET: International Journal of Research in Engineering and Technology, Volume: 03, Jun-2014.

United Scholars
Publications

Email: info@unitedscholars.net
www.unitedscholars.net